Protected Landscapes

IUCN – The World Conservation Union

There are thousands of environmental organizations around the world – but the World Conservation Union is unique. It is the only one whose membership includes both governments and non-governmental organizations, providing a means to work together to achieve effective conservation action on the ground. IUCN's mission – to provide leadership and promote a common approach for the World Conservation Movement – translates into a practical aim to establish links between development and the environment that will provide a lasting improvement in the quality of life for people all over the world.

The Union's work is increasingly decentralized and is carried forward by an expanding network of regional and country offices, located principally in developing countries and working in close partnership with headquarters. Specialist programmes, covering themes such as ecology, species conservation, habitat and natural resource management, and the relationship between population and the environment, draw on scientific input from around the world. They develop strategies and services to address today's critical international environment and development issues – and seek to anticipate and address those of tomorrow. This work is further strengthened via a network of six Commissions, drawing together dedicated and highly qualified volunteers into project teams and action groups – for results.

Parks Congress

A World Congress on National Parks and Protected Areas has been held each decade since 1962, the most recent being the IVth Congress held in Caracas, Venezuela in 1992. The objective of the Congress process is to promote the protection and most effective management of the world's natural habitats and landscapes. This is to ensure that they can make the maximum possible contribution to sustaining human society and maintaining biological diversity.

The IVth World Congress, under the theme 'Parks for Life', reached out beyond professionals directly connected with protected areas to management agencies, non-governmental conservation organizations, traditional people's groups, relevant industries and resource managers.

Specific objectives of the IVth Congress included demonstrating that protected areas can be focal points for more broadly based rural development initiatives; promoting the concept of regional planning as a means of conserving biological diversity; greatly expanding the constituency for protected areas by identifying productive partnerships; expanding the global network of protected areas and further developing a system for monitoring their status; strengthening the application of science to management issues; developing improved concepts of protected areas in coastal and marine habitats; and greatly increasing international support for protected areas. In sum, demonstrating the value of protected areas within wider strategies for the conservation and sustainable management of the earth's natural resources.

Protected Landscapes

A guide for policy-makers and planners

P. H. C. Lucas

Chair, IUCN Commission on National Parks and Protected Areas
Fellow, Environment and Policy Institute, East–West Center

IUCN – The World Conservation Union
Gland, Switzerland

with the support of

The Countryside Commission,
Cheltenham, United Kingdom

and

The East–West Center,
Honolulu, Hawaii, USA

CHAPMAN & HALL
University and Professional Division
London · New York · Tokyo · Melbourne · Madras

Published by Chapman & Hall, 2–6 Boundary Row, London SE1 8HN

Chapman & Hall, 2–6 Boundary Row, London SE1 8HN, UK

Blackie Academic & Professional, Wester Cleddens Road, Bishopbriggs, Glasgow G64 2NZ, UK

Chapman & Hall, 29 West 35th Street, New York NY10001, USA

Chapman & Hall Japan, Thomson Publishing Japan, Hirakawacho Nemoto Building, 6F, 1-7-11 Hirakawa-cho, Chiyoda-ku, Tokyo 102, Japan

Chapman & Hall Australia, Thomas Nelson Australia, 102 Dodds Street, South Melbourne, Victoria 3205, Australia

Chapman & Hall India, R. Seshadri, 32 Second Main Road, CIT East, Madras 600 035, India

Prepared and published with the assistance of the Countryside Commission
First edition 1992

Typeset in $9\frac{1}{2}/10\frac{1}{2}$pt Century Old Style by
Columns Design and Production Services Ltd, Reading
Printed in Great Britain by
St. Edmundsbury Press, Bury St. Edmunds, Suffolk

ISBN 0 412 45530 7

Citation: Lucas, P.H.C. (1992) *Protected Landscapes.*
Chapman & Hall, London. xvi + 282 + index pp.

Also available from IUCN Publications Services Unit, 219c Huntingdon Road, Cambridge, CB3 0DL, UK.

A catalogue record for this book is available from the British Library

Library of Congress Cataloging-in-Publication data available

Contents

List of boxes

Preface

This guide is one of a series of publications prepared as a consequence of growing international interest in the concept of Protected Landscapes and Seascapes.

The cooperation of IUCN with the Countryside Commission, the Countryside Commission for Scotland, and the Council of Europe led to the holding of the International Symposium on Protected Landscapes in the Lake District, United Kingdom, in October 1987.

In preparation for that symposium, IUCN published two publications which were essentially directories of protected landscapes called *Protected Landscapes: Experience Around the World* and *Protected Landscapes: the United Kingdom Experience.* From the symposium came *Protected Landscapes: the Summary Proceedings of an International Symposium on Protected Landscapes.* Subsequently, the Countryside Commission published in several languages The Lake District Declaration (p. xv) which was a key product of the symposium.

There was a clear call for more in-depth information about the concept and how it might be put into practice. This led to a resolution on the concept being adopted by the General Assembly of IUCN at its 17th Session in San José, Costa Rica, in February 1988. Among other things, this resolution recommended that the Director-General of IUCN should 'encourage IUCN members having experience and expertise in the establishment and management of protected landscapes and seascapes to make such expertise widely available . . .'

This guide is a response to that resolution.

Thanks are also due to Andrew P. Stott (Northern Ireland), Michael Dower and Duncan and Judy Poore (UK), Rob Milne and Norah Mitchell (US National Park Service), Terrence D. Moore (USA), Allen D. Putney (US Virgin Islands), José Nosel (Martinique), Erich Cardich (Peru), Clive Anstey, Robin Gay, Gerald Rowan, and Katherine Walls (Department of Conservation, New Zealand), Glynn Christie, Gary Dyet, Phillip Lissaman, Di Lucas and Kura Pitcher (New Zealand), Rowena Childs, Graeme Kelleher and Peter Quilty (Great Barrier Reef Marine Park Authority, Australia), Mitsuo Uruki (Japan), Zafar Futehally (India), and Rob Malpas (Kenya).

I was encouraged by Professor John Aitchison and Michael Beresford, the Directors of the recently established International Centre for Protected Landscapes at the University College of Wales in Aberystwyth, UK.

Finally, I acknowledge the pioneering efforts of the late Reg Hookway, former director of the Countryside Commission, who introduced me to the protected landscape concept.

While the help of all concerned is greatly appreciated, responsibility for the final text rests with me.

P.H.C. Lucas
Chair
IUCN Commission on National Parks and Protected Areas

Foreword

This publication marks a breakthrough in the way we look at protected areas. For more than a hundred years, thinking and action in this field has been driven by the powerful image of the National Park, pioneered in the United States, and then taken up and adopted around the world.

National Parks are more necessary now than ever before. But they are only one among a family of types of protected areas. Nature reserves, natural monuments and so on all have their part to play. Increasingly though, attention is focusing on the contribution which can be made by another kind of protected area; protected landscape – protected areas in which people live and work, but do so in harmony with nature.

This guide to protected landscapes, prepared by the Chair of the World Conservation Union's Commission on National Parks and Protected Areas, Bing Lucas, explains the concept and gives examples of its application in many parts of the world – it shows that protected landscapes, first consciously recognized as a conservation approach in heavily populated parts of Europe, have a potential relevance and application in many countries, developed and developing. In an ever-more crowded world, they will be more important still in the future.

The Countryside Commission joined with the East–West Center and IUCN to make possible the preparation and publication of this guide. It is our belief that – as a former Director General of IUCN put it – protected landscapes are a concept whose time has come. We hope this guide will spread knowledge of the approach, raise awareness of its value and help, in a practical way, in the establishment and management of many more protected landscapes.

Adrian Phillips
Director General
Countryside Commission
UK

The Lake District Declaration

We, the participants in the International Symposium on Protected Landscapes held in the Lake District, England in October 1987, building upon the unique natural characteristics of the area and the great cultural traditions established there by Wordsworth and Ruskin, BELIEVE that:

• People, in harmonious interactions with nature, have in many parts of the world fashioned landscapes of outstanding value, beauty and interest.

• These landscapes, although often much changed from their natural state, make their own special contribution to the conservation of nature and of biological diversity; for many of the ecosystems they contain have evolved and continue to survive because of human intervention. As large areas of undisturbed land become scarcer because of rapidly rising human populations and intensified land use, these landscapes will greatly increase in importance as repositories of biological richness. Moreover, they can serve as vital buffer zones around more strictly protected areas.

• They preserve the evidence of human history in monuments, buildings and the traces of past land use practices. Their continuing use to provide living space and livelihood for indigenous populations allows traditional ways of life and traditional values to endure and to evolve in harmony with the environment.

• They make an important contribution to the physical and mental health of people subject to the stresses of present day life and they offer beauty, pleasure and recreation to many. They give inspiration to writers and artists. They provide young and old with opportunities to learn about their surroundings and comprehend the cultural diversity of the world.

• Even more important, these landscapes are living models of the sustainable use of the land and natural resources upon which the future of this planet and its people depend. They are working examples of principles set out in the World Conservation Strategy and by the World Commission on Environment and Development. They demonstrate that it is possible to design durable systems of use that provide economic livelihoods, are socially and spiritually satisfying, are in harmony with nature, are aesthetically pleasing and preserve the cultural identity of communities. Moreover good conservation has proved to be good economics.

For these reasons, we also BELIEVE that:

• It is vital to protect such landscapes both for their present value and for the contribution that they will make to spreading the philosophy and

practices of sustainable development over much larger areas of the world.
• There should therefore be universal recognition for this concept of landscape protection; much greater priority should be given to it; and there should be an active exchange of experience between nations.
• These inhabited landscapes are in delicate and dynamic equilibrium; they cannot be allowed to stagnate or fossilize. But change must be guided so that it does not destroy but will indeed increase their inherent values. This means for each protected area a clear definition of objectives, to which land use policies within it should conform is needed. It means also a style of management that is sensitive to ecological and social conditions. This will be possible by building upon spiritual and emotional links to the land and by the operation of flexible systems of graded incentives and controls.

We further BELIEVE that:

- The protection of these landscapes depends upon maintaining within them a vigorous economy and social structure, and a population that is sympathetic to the objectives of conservation. It means working with people at all levels, and especially with those living and working in the area – the people most intimately affected by what happens to it.

We DECLARE, therefore, the following actions to be of vital importance:

- that governments, international organizations, development agencies and non-governmental organizations should recognize the crucial role that such landscapes can play in sustainable development and in the conservation of the cultural and natural heritage of nations; and should develop programmes accordingly.
- that governments should adopt the protection of these landscapes as a part of their public policies for the use of natural resources and provide sufficient funds to make this effective; and that they should use these protected areas as models – 'greenprints' – for the sustainable management of the wider countryside.
- that governments and development agencies should direct funds destined for the support of agriculture or other economic objectives in these areas towards kinds of development that favour conservation.
- that national and international organizations should promote a worldwide exchange of information and experience on the management of such landscapes and should encourage and extend training in this field.

We PLEDGE ourselves to the promotion of these principles and actions.

Foster (1988)

1
Protected landscapes

From the fire tower on Bear Swamp Hill, in Washington Township, Burlington County, New Jersey ... to the south, the view is twice broken slightly – by a lake and by a cranberry bog – but otherwise it goes to the horizon in forest...

The people (of the Pine Barrens) have no difficulty articulating what it is that gives them a special feeling about the landscape they live in; they know that their environment is unusual and they know why they value it...

Of all the schemes that have ever been created for the development of the Pine Barrens, the most exhaustive and expensive one is the proposal for a jetport and a new city ... the largest airport on earth – four times as large as Newark Airport, LaGuardia, and Kennedy put together. As a supersonic jetport, it would serve a third of the United States...

McPhee (1967)

Introduction

But for a decision in the late 1970s taken by United States Federal and New Jersey State authorities, the exceptional landscape of The Pinelands or 'pine barrens' (Plate 1) would today most likely be the site of a jetport with a new city and a whole range of associated developments. The unique character of the region would have been changed irrevocably.

Instead, the gently rolling landscape with extensive forests of pine, oak and cedar broken by wetlands is a protected landscape where some half a million people live year round in a pleasant environment, some maintaining row crop farming and cultivating cranberry and blueberry in traditional fashion.

A cooperative management regime involving Federal, State, and community governments has seen the landscape values of the Pinelands maintained in a manner which has ensured that the region's social and economic needs are met. It has achieved this while seeking to respect the rights of the owners of the two-thirds of The Pinelands which are in private hands.

The harmonious balance of people and nature – of conservation, social and economic values – has thus been maintained in what is legally known as the Pinelands National Reserve.

The Pinelands region is one of a growing number of areas worldwide which demonstrate the value of the protected landscape concept in serving people and nature. Because the concept involves resident populations, its success depends greatly on cooperation and on the mutual commitment of the people and the authorities to the concept and the goals and techniques of management.

Landscape has been described by the International Federation of Landscape Architects (IFLA) as 'the environment we experience ... the interaction of natural resources and people's needs'. This was expanded and reinforced by Duncan and Judy Poore in their writing for the International Symposium on Protected Landscapes held in 1987.

Box 1.1: LANDSCAPE

Many things are encompassed in our understanding of the word landscape: the geological structure of the land, its soils, animals and its vegetation; the pattern of human activity – fields, forests, settlements and local industries – both past and present. It is a matter not only of beauty, of aesthetic appreciation of nature and architecture, but of the whole ecology of an area and the history of its occupation and use by people.

Source: Poore, D. and Poore, J. (1987) *International Symposium on Protected Landcapes.*

In the context in which the term is used here, landscape is an interface between nature and culture, the consequence of human presence in the natural environment and the imprint of the natural environment on the culture and way of life of its residents, past and present. The landscape contains important evidence of past relationships with the land as well as present uses. Landscapes are central to a sense of identity, a sense of place.

The landscape is thus more than a natural setting but one where nature and people are mutually significant influences on what is seen. The natural setting has helped shape the residents, their settlements and way of life and, in turn, the people have placed their stamp on nature to produce, in many situations, landscape which is economically and biologically productive and aesthetically pleasing.

The purpose of the protected landscape concept is to reinforce the positive aspects of this people/nature relationship and avoid or ameliorate negative influences which may destroy or damage the harmony between the people and their environment. At the same time, the concept seeks to provide opportunities for the public to visit and experience the area in ways which enhance the local economy but do not prejudice its natural, cultural and social values.

The protected landscape differs significantly from some of the other categories of protected area recognized by IUCN, specifically because most of the land in protected landscapes is in private or communal ownership or occupation with resident populations going about their daily activities. It is the purpose of this guide to explain the protected landscape concept and encourage policy makers and planners to apply it in a manner appropriate to their situation.

Box 1.2: PROTECTED AREA

Any area of land that has legal measures limiting human use of the plants and animals within that area; includes national parks, game reserves, protected landscapes, multiple-use areas, biosphere reserves, etc.

Source: Adapted from McNeely *et al.* (1990).

It is the combination of outstanding landscape and resident population in harmonious interaction which distinguishes the protected landscape from the national park which remains the best known of the categories of protected area identified by IUCN since Yellowstone was designated the world's first such park in 1872. The laws setting up Yellowstone underline the essential differences between a national park and a protected landscape.

Yellowstone was proclaimed a national park because of the outstanding qualities of a large area of land where nature was dominant. The United States Congress and Senate made their intentions clear when they said that Yellowstone should be 'reserved from settlement, occupancy, or sale . . .' The management mandate given subsequently to the National Park Service was for Yellowstone – and other United States' national parks – to be managed in such a way as to leave their natural qualities unimpaired while providing for the benefit and enjoyment of the people ... as visitors but not residents or private owners.

The inspiration of Yellowstone's establishment saw the national park concept expand slowly at first and then gain momentum until, by the time of Yellowstone's centennial in 1972, there were 1000 national parks worldwide, meeting the definition adopted by IUCN. That definition, in brief, sees a national park as a relatively large natural area not materially altered by human occupation, where human use is for inspirational, educational, cultural and recreational purposes.

Box 1.3: NATIONAL PARK (IUCN CATEGORY II)

A National Park is a relatively large area

- *where one or several ecosystems are not materially altered by human exploitation and occupation, where plant and animal species, geomorphological sites and habitats are of special scientific, educative and recreative interest or which contains a natural landscape of great beauty; and*
- *where the highest competent authority of the country has taken steps to prevent or to eliminate as soon as possible exploitation or occupation in the whole area or to enforce effectively the respect of ecological, geomorphological or aesthetic features which have led to its establishment; and*
- *where visitors are allowed to enter, under special conditions, for inspiration, educative, cultural and recreative purposes.*

Source: IUCN (1985).

The national park concept has served the world and nature well but its application is inevitably limited. By definition, it cannot apply to those landscapes (or seascapes) of special quality where resident populations and their resource use patterns are integral but have materially altered their naturalness.

It is in such situations that IUCN recognizes the parallel and complementary category of protected area – the protected landscape. With people a permanent part of the landscape and day to day activities affecting it, these areas demonstrate durable systems of use that provide economic livelihoods, are socially and spiritually satisfying, preserve the cultural identity of communities, and complement and maintain the special natural qualities of the landscape.

> Box 1.4: PROTECTED LANDSCAPE OR SEASCAPE (IUCN CATEGORY V)
>
> *To maintain nationally significant natural landscapes which are characteristic of the harmonious interaction of People and Land, while providing opportunities for public enjoyment through recreation and tourism within the normal life-style and economic activity of these areas.*
>
> *Source*: Adapted from IUCN (1985).

The protected landscape provides a clear and legitimate alternative to the national park in areas where the presence and impacts of resident populations and private ownership either rule out the Yellowstone-model national park or where it is the very harmony of people and nature which makes for an environment of quality and distinctiveness.

Box 1.5: NATIONAL PARKS AND PROTECTED LANDSCAPES: THE DIFFERENCES

IUCN Cat. II National Parks	*IUCN Cat. V Protected Landscapes*
1. Extensive natural area	*1. Outstanding semi-natural landscapes*
2. Protected from exploitation	*2. In productive use*
3. Protected from occupation	*3. Inhabited*
4. Responsibility of Government	*4. Mainly responsibility of local government*
5. Land publicly owned	*5. Land mainly privately owned*

Source: Phillips (1988).

An allied category of resource management which contributes to conservation and sustainability is the multiple-use management area/ managed resource area (Category VIII). This category includes outstanding landscapes or seascapes, two of which are cited as case studies in this guide – Ngorongoro Conservation Area in Tanzania and the Great Barrier Reef in Australia.

Box 1.6: MULTIPLE-USE MANAGEMENT AREA/MANAGED RESOURCE AREA (IUCN CATEGORY VIII)

To provide for the sustained production of water, timber, wildlife, pasture, and outdoor recreation, with the conservation of nature primarily oriented to the support of economic activities (although specific zones may also be designated within these areas to achieve specific conservation objectives).

Source: IUCN (1985).

The range of categories of protected area IUCN recognizes reflects the thinking of its membership including governments and citizen conservation groups from over 120 countries as well as the expert volunteer network of some 500 people making up its Commission on National Parks and Protected Areas (CNPPA).

The range of categories reflects, too, the varied ways of maintaining the world's living resources for their intrinsic value, for their biological diversity, and as the basis for sustainable management to meet human needs.

Box 1.7: CATEGORIES OF PROTECTED AREAS (SUMMARY)

I. *Scientific Reserve/Strict Nature Reserve*
II. *National Park*
III. *Natural Monument/Natural Landmark*
IV. *Nature Conservation Reserve/Managed Reserve/Wildlife Sanctuary*
V. *Protected Landscape or Seascape*
VI. *Resource Reserve-Interim Conservation Unit*
VII. *Natural Biotic Area/Anthropological Reserve*
VIII. *Multiple-use Management Areas/Managed Resource Areas*

Source: IUCN (1990).

The order of categories does not constitute a hierarchy but reflects, in ascending order, the degree of human use acceptable in each case.

Consequently, Category I, the scientific or strict nature reserve, has the specific purpose of protecting nature and natural processes as undisturbed as possible. This contrasts with the protected area categories at the human use end of the spectrum.

Because the protected landscape is generally extensive in area it may incorporate, within its overall boundaries, other categories such as scientific or strict nature reserves (Category I), natural monuments or landmarks (Category III) and managed nature reserves or wildlife sanctuaries (Category IV). It may even have a core national park (Category II) or serve as a buffer to an adjoining national park, reflecting the balance of people and nature in the protected landscape and the dominance of nature in the adjoining national park.

Protected landscapes have been a key management tool for conservation in Europe for 40 years although in some countries they are still officially called 'national parks' – in Japan and the United Kingdom, for example – in spite of a call by IUCN in 1969 requesting governments to retain this title for areas which have Category II management objectives. The concept is likely to have growing importance in other parts of the world where opportunities to establish protected areas which do not accommodate resident populations are increasingly limited.

Protected landscapes may be established through a variety of means. The protected landscapes of England and Wales ('National Parks, Areas of Outstanding Natural Beauty or Heritage Coasts' under the governing legislation) are examples of areas established as a consequence of national initiative. On the other hand, the protected landscapes of France (known as 'regional nature parks') are products of local initiative with proposals being presented to Paris for approval.

The fact that both the top-down and bottom-up approaches operate successfully demonstrates that there is no single path to success in protected landscapes. However, the fullest possible involvement of the residents is fundamental whatever approach is used.

The experience of Nepal emphasizes the key aspect of involvement of the people who live in areas identified for special protection and illustrates how the accommodation of a range of uses can balance the interests of people and nature.

With national parks already established in the 1970s in the Langtang and Sagarmatha regions of the Himalayas, the Nepalese authorities considered how best to provide for the wise use of the more populous Annapurna region. The population of some 40 000 people concentrated in the valley approaches to the mountains clearly had a major influence on the environment. Additionally, the growth of tourism had led to rapid expansion of villages in non-traditional styles of architecture and had created a substantial demand for construction timber and fuelwood. The success of any management mechanism here clearly depended on a cooperative approach with the people. Consequently, the area is being managed as the 'Annapurna Conservation Area' combining the elements of a protected landscape in the occupied portion and a national park at elevations beyond the villages and their influence. The Annapurna case study elaborates on this.

Drawing from these and other case studies as well as the experience of many countries, this guide aims to make the protected landscape concept better known, to set it in the context of other types of protected area and of global thinking and action on the environment. It goes on to identify the values of protected landscapes and outline approaches to their selection, establishment and management, emphasizing the particular importance of close involvement with local communities.

It seeks to make clear that there is no single way to approach the establishment and management of protected landscapes.

However, it emphasizes that certain principles apply universally. A set of such principles was identified in the Lake District Declaration which was adopted unanimously by the International Symposium on Protected Landscapes held in the United Kingdom in October 1987. That symposium was initiated by the Countryside Commission in cooperation with IUCN and the Council of Europe out of a conviction that wider understanding and application of the protected landscape concept has a significant contribution to make in meeting global environmental problems. This new perspective comes at a time when conservation issues have moved from the concern of scientists and conservation activists to the front line of world politics and news.

The enthusiasm generated by the symposium was taken up by the world conservation community when it met in the 17th session of the General

Assembly of IUCN in San José, Costa Rica in 1988. In a comprehensive resolution (Appendix 1) the General Assembly urged international cooperation to promote the concept of protected landscapes and seascapes.

Box 1.8: CONSERVATION

The management of human use of the biosphere so that it may yield the greatest sustainable benefit to present generations, while maintaining its potential to meet the needs and aspirations of future generations. Thus conservation is positive, embracing preservation, maintenance, sustainable utilization, restoration, and enhancement of the natural environment.

Source: IUCN (1980).

This guide is one way by which the concept is being fostered, in the words of the IUCN resolution, 'to link the conservation of natural resources with economic development, following the basic principles of the World Conservation Strategy and the report of the World Commission on Environment and Development'. This underlines the fact that protected landscapes have the capacity to demonstrate in a practical way how the central goal of the World Conservation Strategy can be achieved – 'the integration of conservation and development to ensure that modifiations to the planet do indeed secure the survival and well-being of all people.'

2
Protected landscapes and sustainability

Endangered earth

'TIME' magazine chose Endangered Earth as the Planet of the Year (1989). The newsmagazine highlighted the key issues endangering earth as global warming, waste, pollution, overpopulation and extinction of species. The magazine called for 'a universal crusade to save the planet' and said that 'unless mankind embraces that cause totally, and without delay, it may have no alternative to the bang of nuclear holocaust or the whimper of slow extinction'.

Growing recognition of these global environmental problems in the 1980s saw numerous strategies developed to guide a sustainable approach to use of the earth's renewable natural resources: strategies in which protected landscapes have an important part.

The World Conservation Strategy published by IUCN in 1980 pointed to the need for action to achieve the goal of sustainable development – development which respects the environment rather than plunders it. The challenge was picked up by the General Assembly of the United Nations in setting up the World Commission on Environment and Development (WCED). The Commission, chaired by Gro Harlem Brundtland, then Prime Minister of Norway, and including other leading political figures, had the task of formulating a global agenda for change. Its report, published in 1987 under the title *Our Common Future*, raised global awareness on issues such as population and human resources, food security, species and ecosystems, energy, industry and the urban challenge. The WCED report prompted a political response which, in turn, is making news headlines.

Sustainability and biological diversity

The World Conservation Strategy focused on the need to maintain essential life-support systems, preserve the genetic wealth contained in nature and ensure the sustainable use of species and ecosystems: all processes fundamental to human survival.

One of the most important ways of protecting species and their habitats is through the establishment of legally protected areas. The WCED report recognized that such areas are not luxuries set up to benefit wealthy tourists and wildlife enthusiasts but are essential elements in the search for sustainablility in all countries.

Such protected areas, established generally on public land and with an emphasis on nature free of overt exploitation, are vital to protect 'biological diversity' – the variety and interrelationships of living things on this planet for their value as raw materials and resources for development and human health and as priceless possessions in their own right.

Box 2.1: BIOLOGICAL DIVERSITY

'Biological diversity', (often shortened to 'biodiversity') encompasses all species of plants, animals, and microorganisms and the ecosystems and ecological processes of which they are parts. It is an umbrella term for the degree of nature's variety, including both the number and frequency of ecosystems, species, or genes in a given assemblage. It is usually considered at three different levels: genetic diversity, species diversity, and ecosystem diversity. Genetic diversity is the sum total of genetic information, contained in the genes of individuals of plants, animals, and microorganisms that inhabit the earth. Species diversity refers to the variety of living organisms on earth. Ecosystem diversity relates to the variety of habitats, biotic communities, and ecological processes in the biosphere, as well as the tremendous diversity within ecosystems in terms of habitat differences and the variety of ecological processes.

Source: Adapted from McNeely *et al.* (1990).

However, concern is increasingly being felt that, in many countries, these types of protected areas are becoming islands of nature in a sea of destructive development. Indeed in some countries, the natural values of such areas are being eroded by pressure from outside which may stem from poaching of their resources for profit or occupation by people from surrounding land for survival. The concern is that, if a country's conservation efforts are devoted to strictly protected areas alone, then conservation in the countryside at large tends to be neglected and both biological and cultural diversity are lost.

Recognizing that human survival depends on maintaining a development regime that is sustainable in its use of renewable resources, public interest is growing in establishing protected areas where people live and work but do so in a manner which leaves an important place for nature. Such sustainable use regimes already exist. The need is to find mechanisms to prevent them from being degraded.

In such situations, 'protected landscapes' can effectively be regarded as rural conservation areas with a key and growing role in contributing to nature and cultural conservation, to the physical, social and spiritual well-being of residents and visitors, as well as to the economic benefit of local and surrounding communities. Similarly, the protected landscape is an appropriate mechanism for managing land surrounding a Category II national park where human occupation is incompatible with management objectives because of the particular values in the park. In such cases, the protected landscape, as well as having values in its own right, can serve as an effective buffer to the national park.

In the process, by serving as models for the sustainable management of the wider countryside, protected landscapes have the potential to contribute to reversing the trend towards deterioration of the global environment. They could serve as 'Greenprints for the Countryside' to use the phrase introduced by MacEwen and MacEwen (1987) in their book of that title.

National parks on the Yellowstone model and nature reserves which, for valid reasons, strictly limit human activity have long held the focus of attention for those concerned with ensuring the preservation of the widest possible range of ecosystems and species in the wild.

It is the realization that preservation of species and habitats is a vital component of living resource conservation and that wise use of natural resources is vital to human survival that has led to a wider perspective emphasizing the interdependence of people and nature. The protected landscape is the epitome of people and nature living in the sort of harmonious relationship which should be the universal pattern for the future.

Protected landscapes are also places where traditional plant cultivars are raised, even though they may be in danger of extinction elsewhere. Conserving these varieties can be an important part of the strategy for conservation.

The protected landscape concept is not, however, a soft option for area's which properly deserve to be managed effectively as Category II national parks. The concept should not be abused and used to justify a lowering of standards in areas which, because of their values and management objectives, should be managed to tighter conditions.

Far from being a soft option, the protected landscape can be a more difficult concept to implement because it involves dealing with people who live and work in the area and own property there, as well as with development agencies whose decisions may impact significantly.

Protected landscape goals can be achieved only through mechanisms which influence in a positive manner how people manage land they own or occupy, how various local authorities exercise their functions in the area, and how the policies and practices of outside agencies and governments impact on the landscape. The potential impacts may range from decisions on individual buildings or farms, through major development projects such as dams or highways, to broad agricultural policies.

A global agenda for change

In calling for a global agenda for change, the WCED (1987) expressed confidence that 'humanity has the ability to make development sustainable – to ensure that it meets the needs of the present without compromising the ability of future generations to meet their own needs.'

It called for change, saying that 'economics and ecology must be completely integrated in decision-making and lawmaking processes not just to protect the environment, but also to protect and promote development'. The importance of mechanisms to integrate the environment and development was stressed as opposed to practices of placing responsibility for environmental matters in institutions with little or no control over agricultural, industrial, urban development, forestry and transportation policies and activities.

In saying that the pursuit of sustainability requires changes in the domestic and international policies of every nation and in identifying a need for protected areas to expand into new concepts, the WCED, in effect, endorsed the protected landscape approach in providing a mechanism for the process of integration to be achieved in areas identifiable by their high natural and cultural values.

The Commission said that 'the historical approach of establishing national parks that are somehow isolated from the greater society has been overtaken by a new approach to conservation of species and ecosystems that can be characterized as 'anticipate and prevent'. This involves adding a new dimension to the now-traditional and yet viable and necessary step of protected areas.' The Commission's view was clear – that development patterns must be compatible with the preservation of the extremely valuable biological diversity of the planet.

The Commission could well have said that the protected landscape concept is designed to do just that on a smaller scale by maintaining resource use patterns where a high degree of harmony already exists between development for human benefit and conservation of natural and cultural values.

The World Commission thus gave endorsement from its global perspective to the evolution in thinking which has seen the expansion of the protected area concept to embrace the protected landscape as part of a spectrum of areas, all contributing in various ways to sustaining human society in harmony with the environment. IUCN's Commission on National Parks and Protected Areas, which coordinates IUCN's work with protected areas, including organizing the once-a-decade World Parks Congresses, has been at the forefront of identifying and fostering the new approaches of which the World Commission wrote.

Caring for the earth: a strategy for sustainable living

The successor to the World Conservation Strategy, prepared in 1990 under this title jointly by IUCN, the United Nations Environment Programme (UNEP) and the World Wide Fund For Nature (WWF) reinforces the role of protected areas in the context of sustainability.

Among its nine principles for sustainable living is to 'conserve the Earth's vitality and diversity' which requires a commitment to:

• conserve life-support systems. These are the ecological processes that keep the planet fit for life. They shape climate, cleanse air and water, regulate water flow, recycle essential elements, create and regenerate soil, and enable ecosystems to renew themselves;

• conserve biodiversity. This includes not only all species of plants, animals and other organisms, but also the range of genetic stocks within each species, and the variety of ecosystems;
• ensure that uses of renewable resources are sustainable. Renewable resources include soil, wild and domesticated organisms, forests, rangelands, cultivated land and the marine and freshwater ecosystems that support fisheries. A use is sustainable if it is within the resource's capacity for renewal.

An action point in the chapter entitled 'Conserving the Earth's Vitality and Diversity' is to 'complete and maintain a comprehensive system of protected areas'.

That the Strategy for Sustainable Living clearly sees a major role for the protected landscape is evident from its discussion of the role of a system of protected areas with its references to modified ecosystems and culturally important landscapes.

Box 2.2: FUNCTIONS AND BENEFITS OF A PROTECTED AREA SYSTEM

A system of protected areas is the core of any programme that seeks to maintain the diversity of ecosystems, species, and wild genetic resources; and to protect the world's great natural areas for their intrinsic, inspirational and recreational values.

A protected area system provides safeguards for:

• *natural and modified ecosystems that are essential to maintain life-support services, conserve wild species and areas of particularly high species diversity, protect intrinsic and inspirational values, and support scientific research;*
• *culturally important landscapes (including places that demonstrate harmonious relationships between people and nature), historic monuments and other heritage sites in built-up areas;*
• *sustainable use of wild resources in modified ecosystems;*
• *traditional, sustainable uses of ecosystems in sacred places or traditional sites of harvesting by indigenous peoples;*

►

• *recreational and educational uses of natural, modified and cultivated ecosystems.*

Protected areas can be especially important for development when they:

• *conserve soil and water in zones that are highly erodible if the original vegetation is removed, notably the steep slopes of upper catchments and river banks;*
• *regulate and purify water flow, notably by protecting wetlands and forests;*
• *shield people from natural disasters, such as floods and storm surges, notably by protecting watershed forests, riverine wetlands, coral reefs, mangroves and coastal wetlands;*
•*maintain important natural vegetation on soils of inherently low productivity that would, if transformed, yield little of value to human communities;*
• *maintain wild genetic resources or species important in medicine;*
• *protect species and populations that are highly sensitive to human disturbance;*
• *provide habitat that is critical to harvested, migratory or threatened species for breeding, feeding, or resting;*
• *provide income and employment, notably from tourism.*

Source: Caring for the Earth: A Strategy for Sustainable Living, IUCN/UNEP/WWF (1991).

The protected landscape in context

Traditionally and historically, societies have recognized the value of protecting some areas for their intrinsic worth and so that they can better contribute to the sustainable use of the resources they protect.

For example, South Pacific societies have a tradition of 'tapu' or 'taboo' by which some forest or reef areas are sacred, thus serving as sanctuaries for wild species and ecosystems, often to provide for a sustainable yield of natural resources outside the sacred area.

In other civilizations, the transition from hunter–gatherer societies to a more agriculturally based way of life, saw farming reduce the area of forest and the number of wild animals decline. As Thom (1987) has said, in reviewing the evolution of protected areas in his book *Heritage: The Parks of the People*, 'hunting became the prerogative of the powerful. Hunting reservations were established by kings, and a body of forest law evolved.'

Serving also to conserve the forests for a much wider range of values than sport-hunting, these concepts were the forerunners of today's protected areas.

Information on protected areas has been collected by CNPPA for many years for a wide variety of uses from measuring the adequacy of coverage of the earth's biological diversity to raising public and political awareness. Since 1959, IUCN has also been charged by the United Nations with maintaining a 'United Nations List of National Parks and Equivalent Reserves', work which is carried out under the guidance of CNPPA by the Protected Areas Data Unit (PADU) of the World Conservation Monitoring Centre (WCMC), based at Cambridge, UK.

Initially only national parks and nature reserves were listed but, realizing that this was too narrow a focus, IUCN in 1978 provided an extended framework of categories used for subsequent UN Lists (Appendix 2). These categories recognized that there are legitimate means, other than national parks and nature reserves, of pursuing the goals of nature conservation, particularly in delivering the benefits of nature conservation to local populations.

Specifically, it led to international recognition of the Category V Protected Landscape or Seascape, as well as the closely allied Category VIII Multiple-use Management Area/Managed Resource Area.

The UN List was also expanded to include what the publication describes as areas 'of Special International Significance'. These are Biosphere Reserves, World Heritage Sites (Natural), and Wetlands of International Importance designated by the contracting parties under the Convention on Wetlands of International Importance known as the Ramsar Convention.

The system of categorization is by management objectives recognizing that many countries use their own nomenclature for protected areas. The ease with which each protected area can be categorized depends, of course, on the extent to which countries identify specific objectives for each area. As has been mentioned, many areas known legally as 'national parks' do not fall within the IUCN definition for Category II and are therefore listed under other categories (including Category V) more appropriate to their management objectives.

However, in spite of the increased recognition of protected landscapes, senior officers of PADU said in 1987 that 'three protected area types – strict nature reserves, national parks, and wildlife sanctuaries, defined as IUCN Categories I, II and IV, respectively – have attracted most attention globally as regards protected area establishment and management. As a result, protected landscapes and the objectives of their establishment at international level have perhaps remained largely obscured.'

While this comment is valid in terms of international recognition, the role of the protected landscape is clear, locally and nationally, in those countries where the concept has been applied as a valuable means of achieving social and conservation goals. It was recognized in the World Bank's 1987 policy statement on wildlands which incorporated a protected landscape category, while the European Parliament proposals for 'regional parks' to protect rural landscapes within the European Community aimed at the concept of representativeness and saw them as 'instruments for environmentally sensitive integrated rural management and development'.

Against this background came the initiative of the Countryside Commission in cooperation with the Council of Europe and IUCN to organize the International Symposium on Protected Landscapes held in the Lake District.

'A unique event: an international gathering to look at "protected landscapes", the approach to conservation which is represented by the national parks of England and Wales' was how the Chair of the Countryside Commission (Sir Derek Barber) described the symposium. He said that the Countryside Commission, which has overall responsibility for these areas, convened the symposium 'to offer experience as a backcloth against which those from other parts of the world could exchange experience in the protection of landscapes'.

There were participants from 29 countries and significant participation from international organizations: UNESCO, UNEP, WWF, the Council of Europe, the European Federation of Nature and National Parks and the International Federation of Landscape Architects, as well as IUCN.

This clear indication of international interest in protected landscapes was reinforced in a presentation from PADU based on available but most likely incomplete information which showed 514 Category V Protected Landscapes worldwide, totalling 27 363 600 hectares with a mean size of 53 242 hectares. Areas listed by PADU within the protected landscape category carried 46 individual designations with the term 'national park' used in 16 countries, 'nature reserve' in ten, 'nature park' or simply 'park' in seven, 'landscape protected area' in five.

Summing up the PADU presentation, Jeremy Harrison and Dr Zbigniew Karpowicz pointed out that 'while there are many examples of protected landscapes around the world, it is in Europe that the category is most intensively used ...' and that 'in much of the rest of the world protected landscape systems are just evolving'. The fact is that they exist on the ground but have not been given recognition as protected landscapes by IUCN, by governments or national legal systems.

Reinforcing the potential role of protected landscapes, then CNPPA Chair Harold Eidsvik concluded the symposium by saying that the 'challenge is to bring about a change in public awareness so that protected landscapes ... are recognized as a significant tool in the conservation chest'.

The symposium resulted in a number of products designed to meet this challenge including three publications:

Protected Landscapes: Experiences Around the World, prepared by the IUCN Conservation Monitoring Centre

Protected Landscapes: The United Kingdom Experience, prepared by D. and J. Poore for the Countryside Commission, Countryside Commission for Scotland and IUCN.

Summary Proceedings of an International Symposium on Protected Landscapes, Lake District, United Kingdom 1987.

Another key product of the symposium was the adoption of the Lake District Declaration by which participants identified the values of protected landscapes, the principles underlying them and the actions needed to promote them. The symposium also adopted a draft resolution supporting the protected landscapes concept which was sent to IUCN for submission to its 17th General Assembly and subsequently adopted.

This guide to protected landscapes also represents a key initiative of IUCN through CNPPA and in cooperation with the Countryside Commission and supported by the East–West Center to facilitate and encourage wider use of the concept. These organizations recognize the contribution protected landscapes can make to human society in a variety of fields including conserving nature and fostering sustainability.

3
International cooperation for protected landscapes

The resolution of the 17th General Assembly of IUCN meeting at San José, Costa Rica, 1–10 February 1988 (Appendix 1) supported the points made in the Lake District Declaration as to the value of the protected landscape concept, urged IUCN to develop and promote it and recommended that governments, their agencies and international and professional bodies work towards greater application of the experience already available in the establishment and effective management of protected landscapes.

The resolution also urged support for other international efforts such as the Action Plan for Biosphere Reserves which promote effective management of protected landscapes in ways which respond also to the needs and aspirations of resident populations.

It also recommended that the World Heritage Committee be encouraged to adopt the principle that selected protected landscapes possessing significant harmonious associations of cultural and natural features can be considered of outstanding universal value and worthy of inscription on the World Heritage List.

In the light of the IUCN resolution two internationally recognized classifications of land already referred to deserve further comment in relation to protected landscapes: Biosphere Reserves and World Heritage Sites.

Biosphere reserves

The biosphere reserve concept emerged from UNESCO's Man and the Biosphere Programme (MAB) as an international and interdisciplinary research programme to provide the scientific knowledge and trained personnel needed for sound and sustainable management of renewable natural resources. From this concept, in turn, emerged the idea within MAB of a coordinated worldwide network of protected areas – biosphere reserves – serving conservation as well as research and education.

They were seen as having three essential roles in the MAB programme.

1. A conservation role reinforcing the conservation of genetic resources and ecosystems and the maintenance of biological diversity.

2. A logistic role as a network of areas related to MAB field research and monitoring including training and information exchange.
3. A development role, associating environmental protection and development of land resources for research and education.

The theory of the concept is that biosphere reserves would have a strictly protected core area with a buffer zone with controlled non-destructive use and a so-called transition area where traditional resource use, rehabilitation, experiment, research, etc. would take place.

Biosphere reserves must have adequate long-term legal protection and be large enough to be effective conservation units. They must include representative examples of natural biomes and examples of harmonious landscapes resulting from traditional patterns of land use.

The principles are thus precisely those of protected landscapes and, of the areas discussed in this guide, the Pinelands National Reserve in the United States and the Ngorongoro Conservation Area in Tanzania are biosphere reserves.

The MAB Bureau serviced by UNESCO in Paris, France designates biosphere reserves after nominations from countries which are part of the MAB system are reviewed by IUCN and others. The Bureau then has the task of coordinating subsequent information exchange.

The potential usefulness of the biosphere reserve concept has not always been realized in practice for a variety of reasons, only one of which is that there is no funding mechanism associated with the concept. There have been, in some instances, poor linkages between research and the protected area management agency. In others, intended research programmes have not been implemented or results not fed into the system to facilitate information exchange. In some cases, it appears that the biosphere reserve designation has been seen as just a label added to an existing protected area so that a number of designated biosphere reserves are in fact Category II National Parks lacking the human resource use elements inherent in the concept and fundamental to protected landscapes.

There is certainly value in the biosphere reserve concept that has yet to be fully realized and areas managed as protected landscapes can both contribute and gain from sharing in a pool of experience and research under the MAB structure.

World heritage sites

The World Heritage Convention – the Convention concerning the Protection of the World Cultural and Natural Heritage – became operational in 1978 and has over 110 countries which are States parties to it. It aims to identify, list and protect natural and cultural properties of outstanding universal value for the benefit of all people. However, as presently drafted, the World Heritage Convention is not able to recognize the outstanding qualities of landscapes which derive their appeal from the interplay of natural and cultural values. This is a matter which should be addressed when the convention comes up for review in 1992.

The World Heritage Committee, made up of 21 States whose members are elected by the States parties, makes decisions on listing and on requests for support from the World Heritage fund drawn from the contributions of the State parties and from other sources. Financial support from the fund has averaged about US$1 million a year. The Committee also draws up 'Operational Guidelines for the Implementation of the World Heritage Convention' containing the criteria against which nominations are evaluated.

The process of listing is initiated by States parties to the convention submitting to the World Heritage secretariat, UNESCO, Paris, France, nominations of cultural and natural properties under their jurisdiction for inclusion in the World Heritage List. The nominations are evaluated by IUCN in the case of natural area nominations and, in the case of cultural areas, by ICOMOS (the International Council for Monuments and Sites).

While the Convention nominally brings together culture and nature, as already indicated its text does not cope with them in combination. As a result, a nominated site must meet the criterion of 'outstanding universal value' for both its natural value and for its cultural values for it to be inscribed on the World Heritage list under both headings. Of the areas discussed in the case studies, Taishan in China and Ngorongoro in Tanzania are listed as World Heritage sites under both headings. This clearly does not go as far as was envisaged in the IUCN General Assembly resolution, hence the call for the Convention to be reviewed to recognize the dual natural and cultural values of protected landscapes.

4
Values of protected landscapes

Protected landscapes contribute to society in many ways. Their values are such that benefits accrue to those who live and work in them, who visit them, and to the wider world beyond their boundaries.

> Box 4.1: VALUES OF PROTECTED LANDSCAPES
>
> - *Conserving nature and biological diversity*
> - *Buffering more strictly protected areas*
> - *Conserving human history in structures and land-use practices*
> - *Maintaining traditional ways of life*
> - *Offering recreation and inspiration*
> - *Providing education and understanding*
> - *Demonstrating durable systems of use in harmony with nature.*
>
> *Source*: Adapted from the Lake District Declaration (1987).

Conserving nature and biological diversity

The protected landscape conserves nature and biological diversity, not only through areas such as nature reserves which may come within its boundaries, but through ecosystems the broader landscape contains, along with plant and animal species which may have evolved through human intervention. The importance of biological diversity coupled with the impact on undisturbed land of rapidly rising human populations and intensified land use, greatly increases the importance of protected landscapes as repositories of biological richness. Indeed, there is a view that some managed areas may be more diverse than the natural systems they replace.

Buffering more strictly protected areas

Population and associated increases in demand for natural resources mean ever-increasing pressure on more strictly protected areas from neighbouring landowners and occupiers. Because it is virtually impossible to manage such areas without the cooperation or at least the acceptance of the area's neighbours, there is a growing case for buffer or support zones so that core natural areas can survive. As well as being valuable in their own right, protected landscapes can play this buffering or supporting role for more strictly protected areas. For example, the Annapurna Conservation Area in Nepal demonstrates the vital role of a range of zones from intensive use zone to special management in protecting a core wilderness zone.

Conserving human history in buildings and land-use practices

The importance of preserving evidence of the past through visible remains of evolving human cultures and their buildings goes without saying. However, what is less recognized as a valuable heritage is evidence of past land-use practices.

An example is the moorland in Dartmoor (UK) where Mercer (1984) says, 'the remaining real hill-farmers still set out to look at cattle and sheep on the moor . . . pursuing an art and a mystery that is more than 3000 years old. Bronze Age man invented the system; historic men have merely tinkered with it. It is the lightest touch man has applied to a natural ecosystem, but only by continuing the application does he maintain the moorland.' Here, the protected landscape has a unique role in linking the human past with what the visitor or resident sees today, not only through structures but through the human imprint on the land.

Maintaining traditional ways of life

There is great social importance in allowing those who wish to do so to maintain traditional ways of life and values in harmony with the environment. The protected landscape concept can facilitate this by providing a means to control or refuse intrusive developments which may threaten or destroy the social character of regions or communities.

The policies for protected landscapes, 'regional nature parks', in France place strong emphasis on support of traditional lifestyles and on revitalizing traditional industries to meet the interest of modern consumers in products which are not mass produced. Similarly, the Pinelands area in the United

States has maintained industries in natural products such as cranberries and blueberries which are vitally important economically as well as traditionally to local residents.

Offering recreation and inspiration

Clearly recreation in the physical sense is an important value inherent in the protected landscape but so is re-creation in terms of inspirational and spiritual values.

The Peak district in the United Kingdom is an area surrounded by a major concentration of people where the integration of recreation with conservation and use provides a range of varied experiences for visitors from crowded towns. They can experience walking on lonely moorland, climbing, caving, cycling, sailing, gliding, driving for pleasure, photography, painting, nature study or simply enjoy the beauty of the area.

The Lake District, inspiration for the declaration, was, of course, inspiration for writers such as William Wordsworth and John Ruskin, whereas across Asia the sacred mountain of Taishan inspired Chinese emperors, and other visitors from Confucius to Mao Zedung to write of its beauty.

Providing education and understanding

The educational value of protected landscapes is considerable. They illustrate representative examples of landscapes shaped by people and people shaped by the natural environment in which they live. They give a sense of place to people and provide a contrast to the intensively developed landscapes and cityscapes of modern urbanized societies with, hopefully, a transfer of values and a commitment to make those environments as much a blend of people and nature as is practicable.

To the people of the modified landscapes of the Waipa area of New Zealand, the protected forest remnants and surviving wetlands show how modified their landscape is and how vital it is to retain and enhance what little is left that is indigenous to their region.

Similarly, the diversity of lifestyles maintained in many protected landscapes emphasizes the distinctive values inherent in cultural diversity.

Demonstrating durable systems of use in harmony with nature

Protected landscapes should provide for the economic well-being of their residents, preserve the cultural identity of their communities, be socially and spiritually satisfying, and be aesthetically pleasing. They should be living examples of the sustainable use of resources on which the future of the world and its people depend and, as such, demonstrate the principles of conservation strategies at the regional and local levels. The protected landscape fills a distinctive niche in the spectrum of protected areas. It has an important role in its own right as well as complementing other categories. It is, in a sense, a bridge between predominantly natural and heavily modified landscapes. It is, too, a torch lighting the way to the goal of sustainability.

5
Selection of protected landscapes

Introduction

This section of the guide is aimed at assisting those wishing to apply the concept to identify the most appropriate approach applicable to their situation.

The protected landscapes concept, while of potentially broad application, is inextricably tied to the political, administrative and planning structures of the country concerned. Consequently, while the principles and criteria for selection and establishment of protected landscapes will have a common thread, the mechanisms used will vary considerably.

Since protected landscapes include resident populations whose welfare is an important objective, it is essential to their successful establishment to work cooperatively with the resident communities. It is they who have helped shape the character of the landscape and their cooperation is essential to maintain the harmonious interaction which epitomizes the protected landscape.

It is thus vital that the selection of the area and the form of protection and management decided upon have the fullest possible support and cooperation of the local people and that the legal and administrative structure adopted fits the pattern of local, regional and national institutions and processes of decision-making.

To attempt to transfer in total a pattern which is successful in one country to another with different institutions and decision-making processes is likely to lead to failure. However, there is much to learn from the experiences of other countries, especially those with similar social and economic situations, institutional structures and planning mechanisms.

The challenge to the policy maker and planner wishing to introduce or expand the protected landscape concept is to blend a national perspective with local and regional interests and political and administrative structures (see Box 5.1).

Making a start: the ideal or the pragmatic

Accepting the value of the protected landscape concept, there are a variety of ways in which selection of areas can be approached and a start made.

Box 5.1: A NATIONAL SYSTEM: BASIC NEEDS

An effective national system of protected landscapes requires:

- *An effective national agency to give leadership, a management infrastructure, and trained staff.*
- *A national approach to ensure representative coverage.*
- *A policy for the system.*

Ideally, this would:

- *Encourage community initiative in establishing protected landscapes.*
- *Ensure the effective participation of local communities in the design, management and operation of protected landscapes.*
- *Maintain or enhance the sustainable economic return from protected landscapes (especially to local communities), to the extent consistent with their essential functions.*
- *Ensure public involvement in policy making for protected landscapes, by establishing a body of people from outside the protected landscapes agency (but serviced by the agency) to advise government on the policies of the agency and on other policies that affect protected landscapes.*

An ideal approach would be to set up a systematic process by which a countrywide system of representative protected landscapes might be identified from the outset. However, a nationwide system could prove difficult to implement unless there was clear public understanding of the concept and a supportive climate of public opinion.

A possible approach would be for the proponents of protected landscapes to foster a climate of public and political opinion sympathetic to the concept and to work to establish a protected landscape where the community is strongly supportive and where the public at large sees the area as one of significance needing protection from overuse or from some type of development which would change its character in a manner seen as undesirable environmentally and socially.

For example, the Ngorongoro Conservation Area in Tanzania was established in 1959 to protect the area's wildlife and other natural values while providing continuity to the historic use of the area by the Maasai people whose agricultural and pastoral activities were seen as incompatible with the Conservation Area's former status as part of a larger Serengeti National Park. Conservation Area status, a form of protected landscape, has provided a mechanism by which land use by the Maasai can continue but can be influenced away from activities which led to erosion and destruction of wildlife habitat.

As has already been mentioned, the threat of major development changing both the physical and social character of an area created the climate conducive to the formal establishment of another protected landscape: the Pinelands in USA. Here, community, state and national concern at possible intrusive development converged and provided the springboard for a protected landscape planning structure. Similarly, it was a perceived threat from proposals to drill for oil and to mine limestone that triggered the establishment of the Great Barrier Reef Marine Park in Australia.

A clearly identified need for some form of management to counter internal and external pressures also creates a climate for establishment of a protected landscape. The Annapurna case study demonstrates an attempt in Nepal to work with local people to protect the environment and cater for tourism instead of following the approach used in some situations of relocating the local population or allowing tourism to develop in haphazard fashion.

Another approach is typified by the French structure where national legislation permits local initiatives to propose establishment of protected landscapes as a form of regional development while retaining a national quality control.

Once a working example of a protected landscape is in place, it should then be much easier to apply the concept elsewhere in the country in a systematic manner, building on the experience and on the community and public support gained.

In many countries, *de facto* protected landscapes which already exist but lack a nationally and internationally recognized identity could provide a basis on which to build a system. The Waipa County case study from New Zealand is an example of such an area whose recognition in national law would raise its status and should encourage similar initiatives elsewhere, thus building towards a national system. As it stands, relying largely on local planning subject to regular review, it is very vulnerable to political change.

Fostering awareness

Throughout the process of introducing the protected landscape concept, there needs to be a conscious programme to raise public awareness. This could include working through seminars involving people from the range of interests known to be or likely to be sympathetic to the concept to test the feeling and, hopefully, develop enthusiasm and a commitment to foster broader interest.

This would lead logically to exposure to the communities, industry, and conservation and recreation groups concerned and to local government. This should prompt media interest with the opportunity of generating wider awareness and in-depth articles or documentaries on the concept and on the area or areas where it might be applicable in the country concerned.

In the case of indigenous communities, it would be important to follow the appropriate cultural pattern in both promoting the concept and implementing it. This would mean working with community leaders or elders who may see the protected landscape concept as a means to retain their land, their culture and their authority. Obviously, they would give the lead as to the form of consultation with their people and where it might take place.

Essentially, it is a matter of fostering the concept in the place and manner most likely to be both acceptable and successful. The initiative to foster interest may come from a government agency and/or a national protected landscape authority if the concept has government backing. Alternatively, it may come from the grass roots, from local communities and/or local government or from non-government organizations convinced of a need for action.

A pilot area

The growth of a positive climate of opinion would lead logically to action to establish a pilot protected landscape area.

As indicated, this may be done by focusing on an area which has a high public profile, is of outstanding value, beauty and interest, may be under threat of undesirable change, and where political and public opinion is likely to be sympathetic.

The process of identifying the appropriate boundaries and mechanism to achieve its protection may take time but it would be important to identify the desired objective at the outset and to consider boundaries as soon as possible.

If the initiative did not originate from local people, it is essential that they be involved well ahead of any authorizing legislation as the experience gained and identification of techniques most likely to succeed can help produce more acceptable and workable laws.

All this is particularly vital in societies where the traditional approach to resource use has been threatened by change, perhaps as a result of new technology, increasing population or the rise of tourism. In some cases, the justification for the protected landscape concept may simply be to give legal support to maintaining established traditions, with local people working with the decision-makers to find the most appropriate mechanisms.

If the approach of selecting a single area is followed, it is critical to have the purposes and principles clearly established from the outset and to be satisfied that the candidate area fully meets the criteria.

If an area of lower standard were to become the first protected landscape, then that would prejudice expansion to a national system. There are real dangers in lowering standards to introduce the concept as those lower standards could become the basis for later additions to the system.

However, if the approach set out in this guide is followed, then only genuinely significant areas would be chosen as protected landscapes.

Qualities

Four primary qualities are essential for protected landscapes.

1. Nature conservation values, arising both from ecological processes and those induced by human activities.

2. Landscape values, including historical farming and settlement patterns and local vernacular building traditions.
3. Cultural values, such as lifestyles and social traditions.
4. People-nature relationship in harmonious interaction.

Principles

The nature of landscapes worthy of selection as protected landscapes will vary from country to country. Principles for selection could include the following.

The landscape is a product of harmonious interaction of people and nature

Three essentials, people, nature and harmonious interaction, must be evident in combination, with cultural values as a product of these.

An area may have a visually attractive pattern of land use such as is a feature of some intensive horticultural or grain farming areas. However, if areas lack significant natural ecosystems, they do not possess a key component of the 'protected landscape'.

Conversely, the protected landscape concept is not appropriate for an area whose natural values are essentially unmodified but which is subject to some resource pressure from local people. Such an area is more appropriately classified as a Category II national park with the resource pressures handled through effective management.

The landscape is of outstanding value, beauty and interest

This builds on the Category V definition's reference to 'special aesthetic qualities' and to protected landscapes being characterized by 'scenically attractive or aesthetically unique patterns of human settlement'.

A key element here is that of quality – 'outstanding' and 'special' – clearly indicating that a strong quality element is essential. As quality is subjective, selection criteria need careful consideration as is discussed later under the criterion of quality.

The landscape contains elements which contribute to the conservation of nature and of biological diversity

This reinforces the point that candidates for protected landscape status must contribute to nature conservation. In this way, the protected landscape category is complementary to other protected area categories.

Nature reserves and national parks are primarily tools to preserve biological diversity and representative natural ecosystems. The protected landscape is more a mechanism to conserve, in a cultural setting, quasi-natural landscapes but with natural values an essential part of the sum of values protected.

Such landscapes, although often much changed from their natural state, make their own special contribution to the conservation of nature and of biological diversity; for many of the ecosystems they contain have evolved and continue to survive because of human intervention. The point is reinforced in the preamble to 1988 IUCN General Assembly resolution which says that 'landscapes that have been materially altered by human activities often include species and ecosystems that are dependent on such activities'.

However, it is important to emphasize that where there are areas within a larger protected landscape containing virtually unmodified ecosystems, it is appropriate to give these smaller areas specific protection from human interference by a stricter protected area status within the protected landscape. The protected landscape can then serve as a vital buffer zone around the more strictly protected area.

The landscape preserves the evidence of human history in monuments, buildings and traces of past land-use practices

Historic values may be significant elements of protected landscapes providing a visible link between past civilizations and the present and demonstrating aspects of prehistoric and historic life, buildings and land-use practices.

Dartmoor in the United Kingdom goes back into prehistory with Neolithic people having left widely scattered burial chambers while Bronze Age remains in hut circles, pounds, stone rows, circles, and field boundaries in the open moorland indicate extensive occupation during this period. Limited evidence of Iron Age remains suggest that people of this era stayed at the edges of the moor. Similarly, within the Ngorongoro Conservation Area in Tanzania, where the Maasai people live and work as pastoralists, the protected area includes in Olduvai Gorge one of the major sites in the world for research on the evolution of the human species.

The link between the environment and human occupation is also widely evident in building materials used which historically reflect what has been suitable and available from the area concerned. This use of indigenous materials such as local stone in the United Kingdom's Peak district, reinforces the concept of 'harmonious interaction with Nature'.

The landscape's continuing use provides living space and livelihood for resident populations

This underlines the important place of people in the protected landscape and provides an ongoing challenge and opportunity to all involved. The intention is clear: that resident populations should be able to live and work without, on the one hand, being 'paupers in paradise' or, on the other hand, undertaking activities out of character with the physical and social environment. It does not suggest an open door to development but rather implies room to develop compatibly.

The landscape's continuing use allows traditional ways of life and traditional values to endure and evolve in harmony with the environment

This emphasizes the strong cultural element of the protected landscape and its living evidence of the people/nature relationship. It provides a means by which resident populations may maintain their way of life and values, gain economic benefit and evolve traditional patterns harmoniously with nature, typified by the protected landscapes (regional nature parks) of France such as the Normandie–Maine case study.

The IUCN definition of Category V says that 'traditional land use practices associated with agriculture, grazing and fishing are dominant'. However, although they are dominant, it does not mean they are exclusive as the first principle of harmonious interaction remains fundamental to the protected landscape concept. The IUCN definition reinforces this when it says that 'these landscapes may demonstrate certain cultural manifestations such as customs, beliefs, social organisation, or material traits as reflected in land use patterns'.

All this does not imply a 'museum' approach where residents are placed in a time warp but an evolutionary approach based on the patterns, values and structures which evolve over time.

The landscape can provide for inspiration, recreation and tourism

Public use and enjoyment are important elements of the protected landscape concept which can make an important contribution to the physical and mental health of people subject to the stresses of present day life offering inspiration, beauty, pleasure, recreation to many.

Inspiration in terms of both spiritual and artistic values is an important part of many protected landscapes. It certainly applies to Mount Tai (Taishan) in China, a focus for Buddhist and Taoist pilgrimages and the subject of poetry by many visitors. Sacred places are important in other case studies, for example Fuji-Hakone-Izu in Japan and Annapurna in Nepal.

Artistic inspiration and outdoor recreation were key influences in landscape appreciation in the United Kingdom and France. As the Countryside Commission has said, 'Romantic painters and poets in the nineteenth century influenced the way we look at landscape'. Poet William Wordsworth wrote in his guide to the Lake District of visitors who 'deem the district a sort of national property, in which every man has a right and interest who has an eye to perceive and a heart to enjoy'.

This foreshadowed the later upsurge of popular pressure for improved access to the countryside for the millions of people living in industrial towns and cities which was another major influence in the establishment of the UK protected landscapes.

Today, tourism is an important economic component of protected landscapes needing careful management to maintain its benefits while avoiding overuse of natural resources and abuse of cultural and social values.

The landscape offers education on and promotes public understanding of conservation and cultural values

The protected landscape has a strong educational role in demonstrating people/nature relationships and enhancing understanding of natural and cultural values among residents and visitors alike. It provides, as the Lake District Declaration says, 'young and old with opportunities to learn about their surroundings and comprehend the cultural diversity of the world'. It also provides a basis for encouraging visitors to respect both the natural and cultural values of the area.

The landscape provides opportunities for research

Because it is a controlled environment, the protected landscape offers excellent opportunities for research into people/nature relationships. For example, research in Tanzania's Ngorongoro Conservation Area into cultivation and grazing impacts on its natural values and on its ability to sustain these activities has led to changes in management policies. As a biosphere reserve also, Ngorongoro is able to take advantage of information exchange under the MAB Programme.

The landscape is a living model of the sustainable use of the land and natural resources

A protected landscape is a working example of the strategic approach to conservation endorsed by the World Commission on Environment and Development and encouraged by the World Conservation Strategy and a Strategy for Sustainable Living.

A protected landscape should demonstrate that it is possible to design durable systems of use that provide economic livelihoods, are socially and spiritually satisfying, are in harmony with nature, are aesthetically pleasing and preserve the cultural identity of communities.

The great importance of this is that it underlines the protected landscape's relevance to the future as much as to the past. The protected landscape points the way forward.

The landscape is large enough to ensure the integrity of the landscape pattern

This is quoted from the IUCN definition of Protected Landscapes and Seascapes and is implicit in the principles in the Lake District Declaration. It was inherent in John Dower's thinking in his 1945 report of a protected landscape as 'an extensive area of beautiful and relatively wild country . . .'.

To capture the natural and cultural values in a manner which should ensure the maintenance of their character, preserve their biological diversity and provide the visitor with a distinctive sense of place, the protected landscape needs to be relatively large, desirably in terms of tens of thousands of hectares.

The mean size of Category V areas reported to the International Symposium in 1987 was 53 200 hectares, but protected landscapes in two European countries with national systems of protected landscapes show significantly larger areas. France's 'regional nature parks' average 132 000 hectares and the United Kingdom's 'national parks' average 126 000 hectares. Of those listed in the case studies, only two (Taishan in China and Martinique on the Caribbean island of that name) are less than 100 000 hectares while the largest (the Great Barrier Reef region in Australia) is a huge 34 million hectares.

Box 5.2: CHECKLIST FOR SELECTION

1. *Does the area demonstrate harmonious interaction of people and nature?*
2. *Is it of outstanding value, beauty and interest?*
3. *Does it contribute to conservation of nature and biological diversity and provide opportunities for research?*
4. *Does it preserve evidence of human history and use?*
5. *Does it provide living space and livelihood for resident populations/resource users?*
6. *Does it allow traditional ways of life and values to endure and evolve in harmony with the environment?*
7. *Does it provide opportunities for compatible recreation and tourism?*
8. *Does it provide opportunities for education and promotion of public understanding of conservation and cultural values?*
9. *Does it provide a living model of sustainable use of living resources?*
10. *Is it large enough to ensure the integrity of the landscape/seascape pattern?*

Islands and seascapes

Relativity is important and clearly, if the context is that of a small island situation, a much smaller area could be considered provided it had its own integrity and identity and met the other criteria. In the case of a very small island, an atoll, for example, the atoll and surrounding reef and ocean fished traditionally could qualify as an entity as a protected landscape/seascape.

Indeed, in a small island situation, it is not possible to envisage the 'landscape' being viewed other than in the context of the surrounding sea and the use made of its resources. Consequently, the interface of land and sea should be an integral part of a protected landscape/seascape whose core is a small island or a number of them.

Similarly, it would be highly desirable for coastal landscapes to include a marine component, both to buffer the coastal area from adverse developments in the near vicinity of the shore and to complement the landscape's values with the associated marine values and the cultural interest of traditional fishing patterns.

Conversely, it is very difficult to envisage a protected seascape without a terrestrial component whether in the form of coastal land and the adjoining sea, a group of islands and the surrounding sea, or a reef or combination of reef and islands and the surrounding sea. Here, the linkage between the natural systems and the human use of the resources of land and water would form an essential component of the protected area.

Consequently, this guide does not deal separately with 'protected seascapes'. It does, however, cite as a case study the Great Barrier Reef incorporating the Great Barrier Reef Marine Park and small island protected areas which, together, provide an example of a protected landscape/seascape. Reference is also made later to the role of the Torres Strait Treaty in establishing a *de facto* protected landscape/seascape between Papua New Guinea and Australia.

Boundaries

It is desirable, in selecting candidate areas as protected landscapes, to have boundaries which are sensible, logical and identifiable in terms of nature, people and management.

Clearly, boundaries are important in maintaining the integrity of the protected landscape by ensuring that intrusive development does not take place in locations which will impact adversely on the landscape's natural, cultural or visual qualities.

They are also important so that people – both residents and visitors – are aware of the boundaries because of the differing rules which may apply within them. For example, management mechanisms including planning controls, constraints and incentives may apply within the boundaries but not be applicable outside them.

Because administrative boundaries do not always follow logical geographic or physical boundaries, it may not always be practicable to achieve the ideal. However, the guidelines provide a useful approach to the selection of boundaries.

Box 5.3: GUIDELINES FOR BOUNDARIES

- *Boundaries should encompass complete landscape units and ecosystems.*
- *Boundaries should be convenient and practicable for owners and residents of land within and adjacent to the protected landscape and should be decided in consultation with them.*
- *Boundaries should facilitate effective planning and management.*
- *Boundaries should be convenient for access to the public, consistent with the need to maintain the way of life of residents and other cultural and natural values.*
- *Boundaries should be readily identifiable and should desirably follow natural physical features as these are unlikely to change and are more easily understood on the ground than straight line boundaries drawn on maps. Watershed or ridgeline boundaries are preferable to river boundaries which often bisect important habitat or landscape components.*
- *Boundaries should be drawn taking account of existing or potential land uses on adjoining land which may detrimentally affect or dominate the values of the protected landscape.*
- *Boundaries should include any inland water bodies within the protected landscape.*
- *If the protected landscape extends to the coast, it is desirable that its boundaries include coastal, estuarine, reef and offshore areas which are ecologically linked to the relevant land area and where effective management at the land/sea interface requires unified control.*
- *Boundaries must be recorded clearly in writing and on maps including both a topographical map showing physical features and a cadastral map showing land title boundaries and these maps must be readily accessible to local communities.*
- *Boundaries must be adequately marked on the ground, for example, with markers in key locations along roads and other key access points. Carefully designed, these can be useful for identification and aesthetically pleasing especially if they follow a logo or design theme appropriate to and adopted for the protected landscape.*

When a proposal for a protected landscape is promoted, the proposed boundaries should be identified to avoid misunderstandings and to provide the public with the opportunity to comment both on the proposal and on the boundaries.

A good example of this is in the proposal leaflet issued by the Department of the Environment, Northern Ireland for a Ring of Gullion protected landscape, reproduced as Appendix 3.

The criterion of representativeness

A logical goal is to aim for a system of protected landscapes which represents the variety of landscapes characteristic of the country and which demonstrate harmonious interaction of people and nature.

Representativeness is a goal in the selection of protected areas to maintain the fullest possible range of natural ecosystems and, with them, biological diversity. Representativeness is, by extension, a desirable goal for any national system of protected landscapes with a mix of natural and cultural values.

IUCN's approach to representativeness with protected natural areas has been to adopt and encourage a systematic approach and the World Conservation Strategy identified the need for each country to protect 'a complete range of ecosystems representative of the different types of ecosystem' in that country.

To do this, the WCS proposed that 'each country should review its existing system of protected areas and ascertain the extent to which the different kinds of ecosystem in each biogeographical province are protected'. It urged priority to biogeographical provinces with no protected areas, followed by the provinces in which few of the ecosystem types are represented in protected areas. The WCED endorsed this concept of expanding protected areas with the goal of preserving a representative sample of the earth's ecosystems.

The WCS urged that these global classifications should be used together with more detailed national or regional classifications derived from them.

While developed primarily to achieve the representativeness of protected natural areas, the general approach has validity for identifying representative areas containing significant and harmoniously integrated natural and cultural values. For example, a national ecological classification system, designed primarily to identify representative natural ecosystems, could be overlaid with a system for classifying human influence to provide a means of identifying a representative range of landscapes with interrelated natural and cultural values.

Box 5.4: HUMAN INFLUENCE ON LANDSCAPE

Five primary landscape types have been identified in relation to human influence.

(Protected landscapes candidates would be found in the managed landscape type and in the cultivated landscapes with the greatest emphasis on traditional agriculture.)

1. Natural landscape
Without significant human impact.

2. Managed landscape
Pastureland, rangeland, or forests harvested for wood products. Hamlets are present along with corridors associated with communication and harvest. Species diversity may increase or decrease.

3. Cultivated landscape
Agricultural development predominates usually progressing through three stages:

(a) traditional agriculture with scattered, irregularly shaped cultivated fields next to grazed fallow areas, remnant natural areas;
(b) combined traditional and modern agriculture: similar but with homogeneous cultivation on the best soils;
(c) modern agriculture with remnants of traditional agriculture with mainly homogeneous cultivation broken by scattered patches of traditional agriculture and remnant natural areas.

Species diversity drops considerably in the cultivated landscape. Scattered remnant natural ecosystems are species-poor as a result of repeated disturbances of many kinds and because of their isolation, which inhibits recolonization of species following local extinctions.

▶

4. Suburban landscape
A town and country area with a mix of residential areas, commercial centres, cropland, managed vegetation, and natural areas.

5. Urban landscape
Densely built up areas several kilometres across with remnant managed park areas.
Source: Adapted from Forman and Godron (1986).

Approaches to ecological classification in New Zealand and to landscape assessment by the Countryside Commission (UK) illustrate the sorts of approaches which have proved to be useful working tools in selecting key areas for appropriate protection.

The approach of dividing a country or region into ecological/biogeographic landscape regions provides a framework within which the process of selection of key, representative landscapes can be tackled systematically, followed up by more detailed landscape assessments to confirm and reinforce choices.

The key is to find an approach that meets the needs of the country concerned, is understandable both to those who apply it and to those to whom it is applied, where its conclusions can be documented, and where boundaries can be identified clearly to aid public understanding and to avoid possible legal implications because of a lack of clarity about what is inside or outside the protected landscape.

As indicated, beginning with protection of a single area, the concept may then be expanded to a national system identified under the procedure outlined – establishing a national framework and then identifying significant representative landscapes within that framework.

Box 5.5: A NATIONAL FRAMEWORK

1. *Divide the country into regions taking into account a combination of ecological/biogeographic/landscape/cultural characteristics, each broadly*

▶

homogeneous as to soil, topography, climate, biological potential and land-use pattern.
2. *Within each region identify areas of harmonious interaction of people and nature meeting the selection criteria for protected landscapes.*
3. *Assess each such area in terms of uniqueness and the criteria for quality and integrity.*
4. *Select those areas which best meet the criteria to provide the basis for a representative system of protected landscapes.*

New Zealand is a country which has undertaken the more detailed classification suggested in the WCS to identify representative natural areas for protection.

As a first step, the country was divided into natural or ecological districts based on landscape, climate, vegetation and wildlife. Although this was done primarily to facilitate assessing the representativeness of existing protected natural areas, experience is showing that it is also a very useful basis for assessing the cultural aspects of landscape.

In this context, an ecological district has been defined by Simpson *et al.* (1987) as 'a local part of New Zealand where geological, topographical, climatic and biological processes, including the broad cultural pattern, interrelate to produce a characteristic landscape and range of natural biological communities'. This definition recognizes that human culture strongly influences landscape quality, spatial pattern and natural ecosystems while conversely, cultural activities are strongly influenced by the essential ecological features and processes operating in each district.

Thus, the ecological district emerges as an expression of bioregionalism – of the relationship between nature and culture, a relationship defined by Raymond Dasmann as 'a sense of identity or place (which) develops where an individual grows up within a particular province and learns to recognise its flora and fauna, to respond to its climatic regime, to become familiar with its limits'.

In this context, Simpson *et al.* (1987) consider 'the ecological district is a kind of bioregion which serves to strengthen a sense of place among local people by providing information on the distinctive set of qualities of each district and difference from other districts. This understanding is necessary for appreciation and careful management of landscape'.

In the New Zealand approach, it is recognized that while some ecological districts have their own internal diversity they themselves are part of a hierarchy; many of the ecological districts have characteristics which produce a smaller number of ecological regions. In turn, related regions can be combined into a broader framework of ecological provinces in which significant areas can be identified to achieve representativeness.

The latter, broader, level of classification would be the most appropriate for use as a basis for identifying candidates for a system of protected landscapes which would be representative rather than one concentrating, say, on upland areas.

The criterion of quality

The important criterion of quality is discussed here in the context of the long experience the Countryside Commission (UK) has in assessing landscapes for a variety of purposes, including the identification of candidates for designation as 'areas of outstanding natural beauty', which fall into the broad category of protected landscape.

Before discussing the Commission's techniques of landscape assessment, it is important to say that, just as many of the outstanding national parks of the world were established because their quality was obvious, so is the case for many landscapes deserving protection.

However, the concepts discussed here are particularly relevant if the goal is to meet a legal requirement of quality and/or establish a countrywide system of protected landscapes. The degree of sophistication in the approach will vary taking account of the need and the resources available. However, the Commission's approach (1987a) provides a valuable guide to the approach which has evolved and has been applied with success in England and Wales.

Discussing landscape assessment, the Commission says that:

> a landscape is more than the sum of its component parts. The whole or complete scene perceived is something greater than a series of individual

elements, however significant those components may be in the overall composition. The concern is not so much with the detail of topography, historical and archaeological features and natural habitats or individual features such as trees and woodlands, but with the way in which these separate components together make up the whole scene and how people react to that scene.

Nevertheless, the components are important in the evaluation process, particularly in evaluating an area against the principles which should apply to protected landscapes. The Commission paper goes on to identify a range of the components of a landscape.

Box 5.6: COMPONENTS OF A LANDSCAPE

- *Geology, geomorphology, climate, soil type and vegetation which determine the basic character of the landscape*
- *Flora and fauna; the ecosystems present*
- *Archaeological/historical features such as old earthworks, ridge and furrow systems and landscaped parklands and gardens which record the historic development of the landscape*
- *Human settlement features such as buildings and human artefacts including the farmed landscape with its often intricate mixture of cultivated land, woodland, farm buildings and small settlements traditionally constructed in local materials and contributing to the sense of place*
- *Aesthetic features such as the qualities of form and colour which produce visual effects, the length and breadth of views, and, as an official at an inquiry into one area, put it: 'the feel of the wind, the warmth of the sun, the scent of flowers, the glare of snow, or a sudden break in low scudding cloud, all contribute to the perception of landscape'*

Source: Countryside Commission (1987a).

In discussing its approach to landscape assessment, the Commission points to 'a daunting volume and variety of published material' and defines various terms to avoid confusion.

Box 5.7: DEFINITION OF TERMS IN LANDSCAPE ASSESSMENT

- *Landscape assessment is used as an umbrella term to encompass all the different ways of looking at, describing, analysing and evaluating landscape.*
- *Landscape description refers to the portrayal of what a landscape looks like. This can be done by noting the presence of specific components. The overall scene can be described by using geographical or ecological terms and by reflecting personal reactions to a landscape.*
- *Landscape classification is a method of sorting the landscape into different types and can be a tool for landscape description. A classification does not attach any weight or judgement as to the differences between different sorts of landscape.*
- *Landscape analysis breaks a landscape down into its component parts so as to understand how it is made up.*
- *Appraisal and evaluation are words used for processes whereby landscapes are weighed against particular criteria so as to be given a particular value for a particular reason.*
- *Landscape preference refers to the manner in which people react to particular views – the liking for one particular landscape over another.*
- *Landscape appreciation is the term usually applied to the consideration by someone with a trained awareness, particularly in aesthetics.*

Source: Countryside Commission (1987a).

A fundamental difference exists between 'objective' and 'subjective' approaches. The objective approach is based on the intrinsic qualities of the landscape itself – the 'object' of the viewer. The subjective approach reflects the responses of the viewer – the 'subject' doing the viewing.

Objective approaches are often used to measure and quantify the various components in a landscape. The 'statistical' methods of landscape evaluation or landscape classification have often assessed different landscapes by measuring the incidence of particular elements in any one scene or in any unit area of survey. Subjective approaches have seen, on the one hand, exaltation of wild open spaces and, on the other, preference for landscapes which exhibit diversity and a mixture of elements at the human scale, as is typified by a lowland mixed farming scene.

An approach which has been vindicated by political decision after public inquiry combines the 'objective' with the 'subjective' approach. It is a broad, multidimensional approach based on aesthetic taste operating within the context of informed opinion, the trained eye and common sense.

Landscape quality depends on a large number of factors which, in themselves and in the way they interrelate with each other to form the whole, will create different types of landscape. Those, in turn, will be valued differently by different people because of their human associations. There can be no simple formula for assessing landscapes.

The method of landscape assessment used successfully is a three part one with survey, followed by analysis, leading to proposals – all of which is documented in a manner comprehensible to both the assessors and those who use the assessment.

The key point is that, as with the criterion of representativeness, the approach to assessment of landscape quality needs to be systematic. This ensures that proposals for protected landscapes are made on sound bases.

The criterion of integrity

An area which is representative and has qualities which suggest its suitability as a protected landscape may be so vulnerable to adverse change that its future integrity as a functioning system of human use in harmony with nature may make it unsuitable to pursue as a candidate for protection.

Size and vulnerability are important to maintaining an area's integrity. A relatively small area is clearly more vulnerable to the impact of external development than a large one. The development of a major port will clearly affect the integrity of a coastal landscape, as could an industrial complex on the fringe of an inland one, especially if the areas considered worthy of protected landscape status are relatively small.

If the selection process outlined has been followed then it is a fair assumption that areas identified will inherently have integrity but it is important that the values the area possesses are capable of being maintained, either through the landscape's size providing its own buffer from adverse outside impacts or from the existence of planning control mechanisms which give it protection from potential adverse impacts.

Often it is public concern at possible adverse effects on an area's integrity which provides the spur for protection. This is evident from many of the case studies.

It is in large part a desire to maintain the integrity of both landscape and culture that has prompted residents of the various regions of France to themselves move to establish protected landscapes under the regional nature park concept.

In Japan, the opening in 1957 of a national highway to a previously remote peninsula on the island of Hokkaido mobilized local residents to press successfully for the establishment of a protected landscape in the form of the Shiretoko 'national park' and to battle subsequently to protect its integrity from Japan's massive industrial relocation plan of the early 1970s and, later, from clear-felling of some of its forests by the Japan Forest Agency.

Maintenance of integrity – the holistic, undiminished qualities of a landscape – can thus be both a trigger for establishment of protected landscapes and, once established, a goal for ongoing management.

6
Legal measures and administrative structures

This section discusses the possible nature of legislation for protected landscapes. In the process it discusses the administrative framework for them which would be set out in the statute.

Legislation

Specific legislation to authorize establishment and management of protected landscapes is essential if their protection is to be approached rationally and if national and international recognition is to be achieved.

Even if it is intended at the outset to establish only one area, it is preferable that there be general enabling legislation under which a system of protected landscapes may be established rather than create a need for new legislation each time it is desired to establish a new protected landscape. However, a pragmatic approach, accepting that it may be difficult to achieve general enabling legislation, may dictate stand-alone legislation such as was adopted for the US Pinelands National Reserve and Australia's Great Barrier Reef.

The weakness of having no specific national legislation is shown in the vulnerability of the area within the Waipa District, New Zealand, discussed as a case study. Here, the Queen Elizabeth II National Trust, under its own legislation empowering it to act in the interests of conserving 'open space', responded to local landowners who wished to see landscape values on their farms protected. As well as negotiating voluntary conservation covenants on individual farms, the National Trust worked with the then local authority which built provisions sympathetic to the landscape into the district planning scheme. However, the fact that such schemes are reviewable at five-yearly intervals and that subsequent local government restructuring has seen the Waipa County go out of existence demonstrates the vulnerability of the purely bottom-up approach to protected landscapes without any overall national statute.

The key to success is to find a legal and administrative structure that fits the country and that facilitates the forging of partnership between local and national objectives and that serves as a springboard to achieving common goals for protected landscapes.

Consequently, no simple single approach can be laid down and the guide outlines principles and options and cites examples which should be read along with the case studies in Chapter 11 to help identify the approach most likely to be effective in a given situation.

Legal provisions for protected landscapes could be incorporated in legislation providing for other types of protected areas. However, because of the significant differences between protected areas on public land and protected landscapes, where most of the land is in private ownership, it is usually simpler for protected landscapes to be provided for in separate legislation.

Alternatively, legal provisions could be incorporated in general land-use planning legislation. However, the specific nature of provisions needed for protected landscapes may not sit well with general legislation for overall land-use planning. A possible option would be to recognize the protected landscape concept in general planning legislation and handle the detail in a separate protected landscape statute.

Clearly, there are advantages in having national legislation and in its setting the framework for the concept enabling a system of protected landscapes to be developed within that framework.

Such a framework should provide for:

1. clear statements of the purposes and objectives of protected landscapes;
2. administrative mechanisms to bring national and local interests together in establishment and management of protected landscapes;
3. planning and management mechanisms by which the purposes and objectives of protected landscapes can be applied to the particular situations of each area.

Examples of the latter are the management plan as in Poland and many other countries, the French 'charter' or the UK 'park plan'.

Consequent on the substantial private land component and economic development in protected landscapes, legislation needs to provide for:

1. public consultation on establishment and management;
2. partnership mechanisms to marry national and local interests;
3. linkages with general planning legislation and other land use policy areas such as agriculture and forestry;

4. mechanisms which provide for cooperative management and public use of private land through management and access agreements, covenants and incentives.

There is also a need, at the conservation end of the spectrum, to develop mechanisms to link the management of more strictly protected areas within the protected landscape and, at the development end of the spectrum, to endeavour to harmonize government policies which may run counter to the purposes of protected landscapes.

Additionally, underlying law and practice need to be taken into account such as traditional Scandinavian practices of unrestricted access regardless of land ownership, and English Common Law recognition of public rights of way by established use. Other considerations are practices of communal ownership common in the Pacific region and underlying rights of indigenous or traditional owners or resource users.

Legislation should include the:

- generic title by which protected landscapes will be known
- purposes of the legislation and of protected landscapes
- objectives for which areas are to be managed
- procedures for establishment
- administrative structure
- policy, planning, integrating, control and management mechanisms
- financial, staffing and other provisions

The title

Clearly the legislation needs to identify and name the concept which is the subject of the legislation.

Legislation for IUCN Category V Protected Landscapes is most common in Europe where various titles are used to identify the concept. Many include the words 'protected' and 'landscape' in association with such words as 'area', 'zone', 'region' or 'park'.

Other titles in use in Europe include 'national park', 'nature park' and 'nature reserve' all of which, elsewhere in the world, are normally used for protected areas largely free of resident populations and usually apply to land in public rather than private ownership. Consequently, it is highly desirable to follow the advice of IUCN and avoid such titles which serve to create confusion as to ownership and the objectives of establishment and management. Some other titles have been suggested such as 'protected resource area', 'rural conservation area' and 'heritage landscape'.

While no simple title can adequately convey what is a subtly complex concept, the title 'protected landscape' is already in reasonably wide use and appears more descriptive and less potentially confusing than the alternatives.

Purposes

It is desirable that legislation give a clear indication of its purpose.

For example, the relevant United Kingdom legislation states that areas are protected for 'the purpose of preserving and enhancing the natural beauty . . . of extensive tracts of country . . . for the purpose of promoting their enjoyment by the public.' 'Natural beauty' is defined as including not only scenic beauty but also the geological and ecological qualities of the area. Additionally, the legislation says that those exercising functions under the act 'shall . . . have due regard to the needs of agriculture and forestry and to the economic and social interests of rural areas'.

Thus natural and human values are clearly recognized as is the case in Nepal where a 'conservation area' is defined as for the purposes of 'the protection, improvement, and multiple use of natural resources according to principles that will ensure the highest sustainable benefit for present and future generations in terms of aesthetic, natural, cultural, scientific, social and economic values'.

Most statutes authorizing the establishment of protected landscapes state their fundamental purposes as twofold – for protection of their range of values and for public recreation and enjoyment. A common addition is the social and economic well-being of local communities, an aspect strongly emphasized in France, for example, as well as in the UK and Nepal.

As indicated, it is helpful for the purpose clause to give a lead as to the relative size of the area envisaged for protection as in the UK legislation with its reference to 'extensive tracts'.

It may be desirable to include, as a secondary purpose of the legislation, the goal of achieving protection of representative or rare landscapes which in the aggregate give the nation its own recognizable character.

> Box 6.1: A 'PURPOSES' CLAUSE
>
> *The 'purposes' clause of legislation could say, for example, that it was 'for the purpose of preserving, maintaining and enhancing as protected landscapes representative areas of significant size and of such combination of beauty, landscape, natural, historic, cultural, recreational, community and other values that their management to protect and enhance those values for sustainable use and enjoyment for present and future generations is in the national interest.'*

Objectives

To guide managers, residents and the broader community alike, it is essential to specify the objectives of protected landscapes in legislation. Such a provision would expand on the purpose by requiring that protected landscapes be so managed that the values for which they were established are, as far as possible, maintained and enhanced.

Legislation could say that, having regard to the general purpose specified in the legislation, protected landscapes shall be so administered as to achieve:

1. conservation of biological diversity through protection, management and enhancement of indigenous flora and fauna, ecological associations, traditional plant cultivars and animal breeds, and the natural environment and beauty;
2. conservation of geological, archaeological and historic features in monuments, buildings and evidences of past land use practices;
3. maintenance and enhancement of the harmonious interaction between people and nature which fashioned the outstanding value, beauty and interest of the landscape;

4. maintenance of traditional ways of life and traditional values enabling them to evolve in harmony with the environment to maintain aesthetically pleasing and productive landscapes in a manner which takes account of the social and economic well-being of residents, is sustainable and which maintains the cultural identity of communities;
5. opportunities for the public to have access so that they may gain inspiration, enjoyment, recreation, education and other benefits that may be derived from the protected landscape in a manner and to an extent compatible with maintenance and enhancement of the values for which the area was established.

As discussed later, the legislation should make provision for the purposes and objectives to be elaborated in general policy statements and applied to specific protected landscapes through locally-based management structures and area-specific management plans.

Procedures for establishment

Legislation should lay down procedures for the establishment of protected landscapes.

It is preferable for the legislation to lay down a clear requirement for there to be public consultation rather than providing only for formal public notification in newspapers with written objections to be lodged. Clearly, formal public notice of proposals to establish protected landscapes is essential (desirably with invitations for 'comment' rather than 'objection') but, as discussed in Chapter 8, this should be the minimum.

Legislation could provide for an initial statement of intent to consider establishing a protected landscape with an indication of the reasons, the values and the tentative boundaries.

After a round of consultation on the statement of intent, if the proposal is to be proceeded with, a more specific proposal would be the basis for a further round of consultation before a decision is made.

Both phases should be provided for in the protected landscapes legislation which should also provide the opportunity for those who make submissions to be heard in a non-adversarial situation by the authority responsible for promoting the proposal. Again, local and cultural protocol should be followed.

Indeed, some would say, as in France, that legislation should provide for the initiative for establishment to come from the region itself with the role of the State being to award or withdraw the status.

Designation and withdrawal

Designation of an area as a protected landscape should come from a national level and the legislation should specify what authority makes the decision and by what means – State Council or Presidential decree, Ministerial order etc. – depending on the pattern normally followed in the nation or state concerned. It would be logical for the designation to be made on the recommendation of any national protected landscape authority set up under the legislation. The legislation should provide for the decree or order to include the name to be given to the protected landscape, which would desirably be both brief and descriptive. The decree or order should also include a legal description of the boundaries accompanied by a map showing those boundaries.

A similar process for withdrawal of the designation should be provided for, again following a round of public consultation, to ensure that designation and its withdrawal are not taken lightly.

Administrative structures

Legislation for protected landscapes will obviously need to identify the people or organizations responsible for making decisions and policies about protected landscapes and for implementing those decisions and policies. Although designation should come from a national level, this does not mean that management should be from the national level.

A national protected landscapes authority

A basic pattern of administration for protected landscapes nationally would see involvement of three levels – political, policy and executive.

At the political level, the appropriate Minister would have overall political responsibility for the legislation, which could give the Minister power to authorize the issue of statements of intent to establish protected landscapes, to formally establish them, to alter boundaries or disestablish, to give policy directions, to act as the channel for central government funding, and to provide communication and coordination with other portfolios whose policies may affect protected landscapes.

A national protected landscape authority could be established by legislation to operate at a policy level with nationwide responsibilities to apply the legislation. Legislation could provide for such an authority to be appointed by the Minister, possibly after calling for nominations and could lay down its membership structure.

Appropriate membership of the national authority is vital if it is to have the necessary expertise and if it is to have the confidence of those interested in the very broad range of protected landscapes including affected land owners and communities.

Provision for the appointment of people as individuals with high standing in their own right is generally preferable to legislation which provides for representatives of named organizations when members may be constrained and the independence of the body lessened. The right of organizations to make nominations may also make it difficult to achieve, within the membership, a representative range of skills, age, a balanced sex ratio, and to have an adequate geographical spread. Where the interests of indigenous people may be affected, it is essential to ensure their representation.

While it may be simpler and give more impetus to the concept to opt for the establishment of a new body, it would be wise to look first at the possibility of building on an existing one because of the added cost and possible reluctance to establish a new authority.

In either case, the legislation should desirably spell out the range of qualities and skills needed among members; for example, interest or experience in landscape, nature, history, culture, science, conservation, recreation, tourism, agriculture, forestry, and community affairs.

A national authority would normally be serviced by the executive agency responsible to the government for implementing the legislation.

It is, of course, possible to operate with no separate authority at the national level and have the policy role carried out by the political and executive arms. However, unless the major initiation and policy role for protected landscapes is essentially a regional/local one as in France, the nature of the concept makes it highly desirable to have public involvement at the national policy level as it would be difficult for officials alone to gain the cooperation of the people or to give the public confidence that their concerns would be translated into policy.

Powers of a national authority

A national protected landscapes authority could be empowered by legislation to develop proposals for a national system of protected landscapes (or to respond to proposals). It could be given the legal responsibility to recommend to the political level the issue of the statement of intent, guide the public consultation process by hearing and reviewing public comments and submissions and recommending to the political level the establishment or disestablishment of protected landscapes.

The legislation could provide for the national authority to develop, review and approve statements of general policy for management of the protected landscape system, again after a process of public consultation and taking into account any directions given from the political level in terms of the legislation. It could also be given responsibility to review and report to the political level and to the public on the implementation of the concept. It could also be empowered to recommend funding priorities.

An executive agency

Legislation should identify the agency responsible to implement it. This may be through staff appointed to service the national authority or through a government department already responsible for other types of protected areas.

Desirably, the legislation needs to show a clear separation of the policy/advice/review role of the authority and the executive/implementing/operational role of the executive agency but with close links between the two. For example, the chief executive of the implementing agency should attend meetings of the national authority. There are advantages in the chief executive not being a voting member of the authority: this maintains a clear separation and gives the authority the opportunity to carry out its reviews and make decisions in a more independent fashion and be seen to do so.

Examples of different approaches are the Countryside Commission (UK) and the New Zealand Conservation Authority. In the UK case the Commission is the protected landscape authority and has its own staff: in New Zealand, the Department of Conservation services the policy-level Conservation Authority with the Department as the executive body for protected areas.

Box 6.2: A NATIONAL STRUCTURE FOR PROTECTED LANDSCAPES

Minister	*Political decision-making and coordination*
National Authority	*Policy development, application and review* *Develops/reviews proposals* *Reviews management effectiveness* *Recommends funding priorities* *Reports to Minister and public*
Executive Agency	*Implements decisions* *Applies policies* *Coordinates funding proposals* *Services national authority* *Carries out monitoring and research* *Provides advisory services* *Reports to Minister* *(Its regional arms may also act as the executive agency at the regional/protected landscape level)*

Local structure

Legislation for protected landscapes may provide for a local administrative structure in a variety of ways but there are two essential components:

1. a locally-based authority, board or committee involving the range of skills and interests present on the national authority;
2. an executive agency responsible for implementing the concept on the ground.

The body appointed for a specific protected landscape or regional grouping of them would, of course, work within the national legislative and general policy framework. It would be responsible to apply the legislation and policy through management plans, planning decisions, administering incentives and generally providing leadership, encouragement and advice as the management of the protected landscape evolves in close touch with local communities.

The legislation could provide for the supporting administration to be through a regional arm of the national executive agency, thus providing a mirror image of the national approach. Alternatively, for economy and close integration, local administration may be handled through the local government structure. For greater independence and because a number of local government bodies may be involved, however, the protected landscape may be given its own stand-alone administration.

No matter which approach is followed, where legislation sets up a two-tier national/local structure, it needs to provide for a two-way flow of guidance, reporting and advice between the two tiers so that the concept is not dominated by a top-down approach.

Different emphases in the national/local approach are provided for in legislation applying to France on one hand and England and Wales on the other.

Box 6.3: THE FRENCH APPROACH

The national system of protected landscapes with the greatest emphasis on the regional level is the French. At the time when the areas (known as regional nature parks) were being set up in the late 1960s, the administration of France was still highly centralized and the legislation enacted for the protected landscapes was a pioneer effort in decentralization.

▶

Protected landscapes are set up on the initiative of local groupings ('syndicats mixtes') who then present their proposals for approval by the executive in Paris.

The local boards ('syndicats mixtes') are representative of regional and local government, chamber of commerce and industry interests. There is no State representation. The syndicats mixtes provide impetus, co-ordination and moral authority but regulatory power remains with the local authorities which sought the establishment of the protected landscape.

Each protected landscape has a permanent staff, a typical one having sections dealing with ecology, architecture, outdoor recreation, tourism and economic development. The staff's efforts are complemented by non-profit making associations which support the administration in ways ranging from providing a link with the agricultural community and fostering conservation activities to operating local radio stations.

Box 6.4: THE ENGLISH AND WELSH APPROACH

Legislation for England and Wales provides for a more 'top-down' approach with responsibility for identification of areas for special protection lying with the Government-appointed Countryside Commission. Among its powers, the Commission identifies areas as protected landscapes under three titles: National Parks, Areas of Outstanding Natural Beauty (AONBs) and Heritage Coasts.

The mechanisms for local administration vary. In the 'national parks', there are boards/committees with two-thirds local government appointees and one-third central government appointees. These bodies exercise statutory planning powers normally vested in local government and operate under the protected landscape legislation and general policy guidance from the Countryside Commission. The local boards/committees generally work through local authority administration although some have their own. In each case, however, there is a 'National Parks Officer' with prime responsibility for management.

▶

This is a more direct and effective involvement than with the AONBs and Heritage Coasts where statutory and management planning remains with local government with the Countryside Commission having an influential role through policy guidance and grants.

'One-off' protected landscape

In some cases, it may be decided not to legislate for a national system but rather to provide for a specific area of national significance to be managed in this way.

The approach already outlined is equally valid in terms of the need to define the area, the purpose and the objects for which it is to be managed.

It is equally important for the legislation to provide for a coordinating authority. In this situation, however, it desirably should include people with a national perspective and those whose interests are more locally based.

The US Pinelands is one example, dealing with a single protected landscape, with a special-purpose commission mainly of State appointees but with Federal government representation and strong planning powers rather than a management role. The Great Barrier Reef is similar and Tanzania is moving in the same direction with an Ngorongoro Conservation Area Authority representing both national and local interests.

These demonstrate variations in approach and reinforce the need for legislation which provides a structure which is effective and acceptable, blending the national and local interest to achieve protected landscape goals and give resident communities an effective voice. One common denominator, however, is the legislative responsibility on the protected landscape body at the operational level to prepare a management plan or charter for the area concerned and for that body to have either direct responsibility for or strong influence on planning and management decisions in the area.

Protected seascapes

With a protected seascape, generally special arrangements need to be made depending very much on whether or not there are resident populations in the terrestrial component of the seascape. Two examples follow, where different national and state/provincial governments are involved.

Box 6.5: TRANSFRONTIER PROTECTION: THE TORRES STRAIT PROTECTED ZONE

An innovative international approach which has resulted in the establishment of a protected seascape is provided for in the Torres Strait Treaty, an agreement signed in 1978 between the governments of Australia and Papua New Guinea after long discussions between the governments involved and representatives of the Torres Strait Islanders.

The agreement clarifies boundaries between the two countries in terms of both seabed and fisheries jurisdiction and identifies what the Treaty calls a 'Protected Zone' recognized by Australia and Papua New Guinea as needing special attention. The Protected Zone is established and rules laid down so that Torres Strait Islanders and the coastal people of Papua New Guinea can carry on their traditional way of life, moving freely, and ensuring that commercial fishing is in harmony with traditional fishing. The establishment of the zone also helps preserve and protect the land, sea and air of the Torres Strait including the native plant and animal life.

Fisheries officials from both countries handle operational matters and there is a Joint Advisory Council to review how the Treaty is working. The Council has members from each country, including state and provincial governments, and traditional representatives.

Box 6.6: AUSTRALIA'S GREAT BARRIER REEF

The Great Barrier Reef case study shows the way in which legislation caters for a particular situation with the approach adopted driven largely by the need to find a way to bring together the interests of the central government of the Commonwealth of Australia and the State Government of Queensland.

Here, the Marine Park Authority is a high level one representative of those governments and responsible to a Joint Ministerial Council. There is

▶

also a Consultative Committee made up of representatives of government, industry and community bodies. All this is backed by an extensive process of public consultation.

Consistency of management of the Marine Park and the smaller protected areas under Queensland control is achieved by the fact that day-to-day management of both the Marine Park and the Queensland areas lies with the Queensland National Parks and Wildlife Service.

Statutory planning

There are a variety of approaches which may be followed in protected landscapes legislation in relation to existing statutory planning arrangements. The approach to follow depends largely on the strength and effectiveness of general legislation for land-use planning and the extent to which protected landscape goals may be achieved through it.

Some possible approaches are as follows:

1. powers to control land-use and development planning may be vested directly in the body administering the protected landscape. This may be justified in the absence of effective general planning mechanisms; or to give a clear priority to protected landscape values even where there is a general planning structure; or it may be linked to lower level planning by requiring those plans to conform; or because a separation of planning functions may be artificial; or
2. powers to control planning may be exercised through the general planning statute but with decision-making vested in a joint protected landscape body representative of both national and local interest; or
3. planning powers may be left with the local government structure with statutory guidance for it to have regard to protected landscape values and/or simply a moral persuasion influence from the protected landscape authority, with a mechanism for consultation. This approach is more likely to succeed where the move to establish the protected landscape comes from local rather than national initiative.

However, the purpose of a protected landscape is most likely to be achieved where there is a capacity to influence change in both urban and rural areas. In the built environment, this means an ability to constrain or influence expansion or redevelopment of urban communities and in rural areas to control afforestation, significant changes in farming patterns and any other incompatible development.

Policy

Whatever approach is adopted, the protected area legislation should require the relevant protected landscape authorities to develop policies – the national authority developing a general policy for the system of protected landscapes and the local protected landscape body applying that policy to the relevant protected landscapes in the form of a management plan or charter. The legislation should require that both the general policy and the specific plans should be drawn up through a process of public consultation.

Control and management mechanisms

The protected landscape legislation should give the specific powers and controls considered necessary for effective protection of the landscape that are not provided for in other legislation.

While there will inevitably be a need to be able to impose some controls or prohibitions, the nature of the protected landscape concept is that the legislation should authorize voluntary arrangements such as convenants and management agreements for protection and management of private land, including access over it. The legislation should also authorize incentives such as grants to achieve protected landscape objectives.

It may also be necessary to have procedures where there are compulsory powers to safeguard critical habitats or access, for example. However, because of the importance of community cooperation, these provisions should be kept to a minimum and should be subject to review.

The importance of education programmes, incentives, and fostering rural conservation projects is such that these should be considered among the statutory role given to those responsible for protected landscapes.

Box 6.7: CONTROLS AND INCENTIVES

Typically, laws governing protected landscapes in Europe prohibit or control hunting, the removal of indigenous plants, mineral extraction, construction of earthworks or buildings or conversion of moorlands or natural grassland.

Laws for the Ngorongoro Conservation Area give priority to resource and wildlife conservation, rangeland and livestock management, and to control development of tourism and rural development for the benefit of the resident population to improve their quality of life and lessen pressure on the area's resources.

Nepal's Annapurna Conservation Area relies on the ability to recycle visitor entry fees into reforestation/firewood projects and village improvements coupled with community education programmes to restore resource conservation practices and to encourage a return to traditional building styles.

Financial and other provisions

Stable funding is critical to successful implementation of the protected landscape concept and, while legislation cannot guarantee this, it may be drawn to provide for a formula for cost-sharing between central and local government for both capital and ongoing costs.

The ability to generate revenue and retain it for recycling into projects of benefit to the area and its residents is an important aspect which desirably should be authorized by legislation. Recreation and tourism are obvious targets for revenue raising for it is the effective management of a protected landscape which makes it an attractive place to visit. Revenue may be generated in a variety of ways depending on the situation, practicality and other factors. Entry charges, bed levies, commercial operating rights or concessions are examples.

In addition to authority for expenditure on grants and incentives to landowners and provision for relief on land taxes, etc., legislation needs to have the usual provisions to finance and facilitate the operation of the administrative structure covering its staffing and operations, including the ability to enter into contracts, agreements and to develop appropriate public facilities.

Staffing

While it is possible to achieve a degree of progress with protected landscape establishment and implementation through the existing structure of central and/or local government, greater impetus will result from there being legislative power to appoint staff dedicated to the task nationally and locally.

The protected landscape's chief officer, manager or director would have statutory responsibility to implement the legislation. He or she would, of course, need to be supported by such other professional, field and administrative staff as the task warrants.

Complementary law and legal arrangements

In addition to the specific legislation to establish and manage protected landscapes, there will generally be a variety of other legislative and legal mechanisms which can support the concept in significant ways.

The main areas of law which facilitate protection of protected landscape values are:

1. land-use planning and control;
2. laws relating to agriculture and forestry and conferring special status on certain types of landscape and key components in the landscape;
3. arrangements for voluntary protection of landscape.

The extent of these provisions and their effectiveness in law and in practice vary greatly but some examples are discussed in Chapter 7 to make clear their real and potential importance in landscape protection. The impact of taxation laws is also discussed there.

7
National policies and protected landscapes

The political dimension

Decisions made by national governments on major policies or projects and by groups of governments working together on such matters as agricultural and trade policies can impact greatly on protected landscapes. In such situations, it is important that the Minister responsible for protected landscapes uses his or her influence to guide such policy-making and development proposals into decisions which will be beneficial to protected landscapes.

An obvious example of this is the move in the European Community away from subsidies for increased agricultural production to support a return to traditional methods of agriculture. Another is the move in the United Kingdom away from tax incentives for forestry development which had led to substantial exotic plantings.

While the local management structure may work hard to influence these types of activities, the greatest opportunity to influence them must come at the national level. The national protected landscape authority has a key role here, desirably backed by the strength of public opinion and the support of sympathetic non-government organizations. Clearly, this demands a flow of information both ways so that the protected landscape administrative structure can respond to proposals at an early stage and, if necessary, seek to influence broad policies or major projects in a manner which recognizes the status and particular values of protected landscapes.

Government agencies may also be large landowners inside a protected landscape and their plans for those lands may or may not be compatible with the concept. Similarly, government agencies may have policies which can be very influential and their activities or plans may or may not be compatible with the goals of protected landscapes. This can apply in the fields of agriculture, forestry, water, energy, regional development, transport, communication and defence.

Decisions taken by government or government agencies in these fields may greatly influence the work of the protected landscape administration but may be legally beyond that administration's control. Consequently, the influence of the Minister responsible for protected landscapes in a manner beneficial to the concept is vital.

Some discussion follows on broad policies of governments which can have a significant impact on protected landscapes – an impact which may be either negative or positive.

Taxation

While taxation policies have been used to encourage increased rural development in agriculture and forestry, often with adverse effects on the landscape, tax mechanisms are also used to encourage action which is beneficial to the landscape.

Relief on land tax may be given under New Zealand law to encourage protection of indigenous forest remnants and wetlands by excluding these areas from property valuations on which taxes are based. Another approach adopted in the United Kingdom is to reduce estate or death duties and charges of various sorts on land identified for protection, where that protection is legally assured.

Land-use planning and control

The machinery elements of land-use planning in relation to protected landscapes have already been discussed but it is appropriate to consider the broader issue of such planning because of its importance to the concept.

The proposition that the State has a legitimate interest in imposing limitations on the use of land within its jurisdiction and that it should have power to implement those limitations is now broadly accepted. As a result, many countries have established a system of land-use controls, although the extent and application vary widely.

Some States have introduced a very strict system which requires, in principle, that consent be obtained for any substantial development activity such as building, engineering, mining or other operations in, on, over or under any land.

Other jurisdictions have contented themselves with controlling development through a zoning system which usually empowers local government to draw up a scheme for the area under its jurisdiction by which each part of that area is allocated for various permitted uses – residential, commercial, industrial, open space, rural, etc. Each area for which a set of permitted uses is designated is a 'zone' and, usually, within that zone variations to the permitted use will be allowed only if they are ancillary to the permitted uses and subject to conditions. For each zone, supplementary regulations or ordinances may be drawn up limiting, for example, the density of buildings, their maximum height, etc. While this system seeks to exercise a general control over development to preserve the overall character of a locality, it does not seek to impose an individual control on each activity.

Such systems have obvious benefit for landscape protection and often dovetail or are legislatively linked with protected landscape legislation to ensure that protected area values are given specific recognition in decisions made under planning laws.

However, it is rare to find a planning control system that extends to all areas and governs all activities which are capable of influencing or damaging landscape values. In many planning systems, important landscape influences remain outside the scope of planning controls, notably agriculture and forestry.

In some countries, planning control is at present regarded as primarily a tool for use in urban areas but, in others, the land use control system has developed special features to protect components of the landscape. Some countries have planning codes which require specific consent to be obtained before woodlands or groups of trees may be felled or even severely pruned and these codes may provide for replanting with similar species if removal is approved.

At times, difficulties have arisen as to whether or not control on the aesthetic impact of development activities can properly be attempted by the planning control system. Objections to planning control being exercised in this field are usually based on the assertion that judgements of this sort involve matters of taste which are too subjective for administrative decision and that their operation amounts to an unwarranted interference with private property rights. However, it is now generally accepted that aesthetics (including landscape values) may form a proper basis for planning decisions. This applies even in jurisdictions such as the United States which are least tolerant of the suppression of private rights in the public interest.

There are examples of a legislative requirement for landscape values to be integrated into the general planning. In Japan, the National Land Use Planning Act requires that areas of unsurpassed or exceptional natural beauty should be classified in prefectural land-use plans. Switzerland is another country where land-use controls may be employed to preserve landscape values. A 1979 federal law obliges each canton to integrate into its development plan, areas which should be preserved because of their scientific or habitat values.

Box 7.1: VISUAL QUALITY PLANNING CONTROL IN NEW ZEALAND

A number of district planning schemes in rural areas of New Zealand's South Island set out specifically to protect landscape qualities.

The Akaroa County District Scheme covers a hilly peninsula. The scheme objectives recognize its important landscape qualities: volcanic skyline, open pastoral ridges, wooded gullies. Buildings and forestry uses are permitted but subject to performance criteria. These criteria focus on key visual features. The height of a building, physical location and visual impacts. In the case of forestry, a management plan is required for proposals in visually sensitive areas. These have been identified on a map which forms part of the district planning scheme. In addition, the scheme includes landscape guidelines. These assist developers to identify specific design criteria and to produce proposals in harmony with the criteria.

The Hurunui County District Scheme recognizes the landscape character as an important resource, both natural and human-influenced. Land-use policies for development seek to safeguard the visual quality by requiring compatibility of proposed development and land-use changes, with the existing visual character and expressions of form, line, colour, texture and pattern in the landscape. In the field of commercial forestry, for example, proposals of over 20 hectares are conditional uses requiring a planning application which gives third-party objection rights. Developers are required to submit management plans to demonstrate conformity with the visual character before their proposals will be considered.

The Malvern County District Scheme identifies particular alpine areas as scenic corridors. The boundaries of these corridors are defined on the

▶

basis of visibility from the Trans-Alpine Railway and State highway connecting the west and east coasts of the South Island. 'Extensive grazing' and 'protected areas' are automatically permitted uses. Building and forestry proposals are controlled uses. The particular features of concern are design, appearance and location of proposals. Landscape guidelines are available to provide guidance on how proposals can enhance the visual qualities rather than prejudice them.

Source: Adapted from G. D. Christie, Works Consultancy Services, Christchurch, New Zealand.

A major failing of some development control systems is that, while they can deal effectively with small projects, they may not be as effective with major projects.

These shortcomings are being met increasingly by requirements for environmental impact assessments such as are required under the US National Environmental Policy Act 1969. Here, one of the triggers which initiates an impact assessment is the effect of the proposed development on scenic and amenity resources. Similarly, a 1985 European Community Directive requires the preparation of impact assessments for a long list of projects and includes the effects of the proposals on landscape among the matters to be considered in assessments.

This approach can cater for major central government proposals which would normally take precedence over local planning mechanisms and be beyond the jurisdiction of a protected landscape administration but which may have the potential to dominate and destroy physical and aesthetic values of the landscape.

Agriculture and forestry

Most agricultural practices lie outside the scope of planning control despite their potential impact on landscapes, for example, by clearing forests and hedgerows, draining wetlands and by intensive development of natural grassland and moorland.

Many States, through taxation, agricultural, and forestry policies have offered incentives hoping to stimulate agriculture and forestry and this has encouraged the adoption of what are often environmentally undesirable practices, harmful to conservation and landscape values.

Moves to 'rationalize' farming by consolidation of small agricultural units leading to the development of large fields adapted to mechanized farming methods is having a profound effect on landscapes in Europe and North America. This has led to France, for example, subjecting such schemes to environmental impact assessment. However, international concern at surpluses of some farm produce and, in some States, a trend to reduce or eliminate farm subsidies is altering this situation. For example, the European Community in 1985 began to encourage member States to introduce national schemes to promote agricultural practices compatible with the conservation of natural habitats in 'Environmentally Sensitive Areas'. In such areas, subsidies are now available to encourage extensive rather than intensive agriculture.

In many countries, the felling of forests is subject to a permit requirement even when the forest is on private land. Although the main purpose of this is to ensure orderly use of forest resources, it has considerable potential benefit for landscape values. Although approvals to fell timber often require replanting, there is often no specification as to species so that exotics can be replanted with adverse effect on landscape quality. In some Spanish provinces, however, replanting must be in indigenous species and in Maryland, USA, where the natural pine cover is removed it must be replanted by the same species.

Some States have controls on planting of areas not previously forested and, in some cases, in both planting and clear-cutting, foresters may be obliged to ensure that boundaries follow natural contours of the land.

Where international and national policies toward agricultural and forestry development prejudicially affect protected landscape values, the battles in support of those values must be fought at the political and national level. However, there are mechanisms at the protected landscape level which can be used to ameliorate the adverse affects of such policies. These are discussed in Chapter 9 of this guide.

Coastal zone and wetlands

The coastal zone is a critically important one in many protected landscapes and the absence of broad planning controls has seen the values of many coastal landscapes compromised.

Industrial development on the coast may result in unsightly structures and poorly treated effluent, reclamations and sea walls may have impacts beyond their immediate surrounds, marina developments may provide facilities for a minority but compromise access and enjoyment for the majority. Similarly, wetlands which may be biologically rich and visually a key element in the landscape may be prime targets for drainage and subsequent development.

Special provisions exist in the planning systems of some countries to protect the environmental values of coastal areas as well as rivers, lakes and wetlands and their surrounds. Such measures have obvious landscape protection benefits by controlling or restricting activities and protecting areas or specific habitats. These may cover a range of purposes from nature conservation and water quality, public recreation, access and navigation.

National policies which recognize the natural values of the landscape and ecosystems and direct local government to plan for wise use of coastal areas and catchments can create a very positive climate for protected landscapes.

Box 7.2: GUIDELINES FOR COASTAL ZONE AND WETLANDS POLICY

Coastal management may be required to conform to a national policy for the coastal zone with local government to prepare and implement conforming coastal zone policies according to the following priorities.

1. *Protection of significant conservation values.*
2. *Maintenance and restoration of the essential character and functioning of each environment.*
3. *Preservation of estuaries, and of predominantly unmodified islands, reefs, coastal wetlands, lakes, lagoons, ponds, and dunes.*
4. *Restoration of degraded water quality.*
5. *Prevention of any new discharges of untreated waste into water.*
6. *Recognition of interests of indigenous people.*

▶

7. *Maintenance and improvement of public access to and along the coast, and of opportunities for recreation which would neither modify the environment nor adversely affect the enjoyment of other users.*
8. *Prevention of the alienation of foreshore, seabed, and public lands immediately adjacent to the foreshore.*

Implementation would mean planning and decision-making requiring:

1. *Coastal developments to conform to a national policy for the coastal zone and for regional government to prepare and implement conforming regional coastal zone plans;*
2. *Specific consents after environmental impact assessments to proposals for reclamations, marinas, non-traditional marine farming, structures such as breakwaters, wharves, jetties, slipways and boatsheds;*
3. *Control or prohibition on removal of sand or reef material, disturbance of coastal vegetation such as mangroves, seaweed, and disturbance of nesting places and migration paths of species such as marine turtles and penguins;*
4. *Constraints on siting of residential, tourism and industrial enterprises on or near the coast including controls on discharge of effluent into the sea or other water bodies and the extraction/use of water;*
5. *Guarantee of public access to and along the coastline, lakeshores or riverbanks;*
6. *Management of wetland catchments so that the complex relationships between the wetland and surrounding ecosystems are taken into account;*
7. *Protection of specific areas of the coastal marine environment for their natural values and providing integrated protection of the land/sea interface.*

8
Protected landscape policy and partnership

Introduction

It is essential for all involved with protected landscapes to have a clear and common understanding of the concept, the goals and the mechanisms by which the goals are to be achieved.

This involves the development through public consultation of general policy statements by the national authority and then effective communication of these to all interested organizations and people.

Implementing these policies then involves partnership and cooperation between the national protected landscape authority and their local equivalents and between the authorities and their executive agencies.

Similarly, at the operational level, it is essential that there be effective communication and positive working relationships with local government, community organizations and individual landowners, occupiers and residents in and around the protected landscape.

A general policy for protected landscapes

While legislation should set out clearly the purpose and objectives of protected landscapes, these will need elaboration in statements of general policy. These statements should elaborate on how the legislation is to be applied through management plans which relate the general policy to the particular situation and needs of specific protected landscapes. This is essential so that while there is a consistent overall policy nationally, that policy is applied to individual protected landscapes in a manner which takes account of regional differences and the specific situation of each area in the national system. The preparation of such a general policy would be a prime responsibility of the national authority working through its executive agency and reviewing responses to a draft general policy from the public and interested bodies before adoption of the policy.

The adoption and publication of policy statements and their ready availability to all interested should greatly facilitate understanding of the goals of protected landscapes and the approach proposed to be followed to achieve them.

Box 8.1: GENERAL POLICY FOR A PROTECTED LANDSCAPE SYSTEM

A general policy for a system of protected landscapes may include such headings as:

1. *Purposes and objectives elaborating on the legislation.*
2. *Application of general policy setting out the role of the policy statements and how they should be applied. For example, it would need to point out that the statements are a guide to action which will depend on local situations and that management plans for each protected area will reflect these variations. However, the policy statements should point out that the overall aim is to ensure consistency so that national objectives are not compromised by local practices.*
3. *Policy development. A brief section could say how statements of policy are developed through consultation etc. and outline procedures to be followed to vary and/or review them.*
4. *Implementation. This should state whose responsibility it is to ensure that the general policy is implemented and should outline the administrative structure for its implementation.*
5. *Policies. These should cover statements dealing with the range of matters considered relevant to protected landscapes including:*
   - *(a) Representativeness*
   - *(b) Size and boundaries*
   - *(c) Relationship with adjacent land*
   - *(d) Management plans*
   - *(e) Research and monitoring*
   - *(f) Conservation of the biological environment*
   - *(g) Conservation of the historical and cultural environment*
   - *(h) Relationship with agriculture and forestry*
   - *(i) Policy on significant developments (such as buildings, roads, industry)*
   - *(j) Social and economic development*
   - *(k) Relationship with local communities and businesses*

▶

(l) *Visitor use, recreation and tourism*
(m) *Visitor safety*
(n) *Signs and information services*
(o) *Conservation education and interpretation*
(p) *Commercial operations associated with visitor use*
(q) *Consultation*

From time to time there will be a need to develop other complementary policy statements on specific matters such as future policy priorities. Clearly, too, there will be a need to review the general policy in the light of changing circumstances, increased knowledge and the experience of those responsible for applying the policies at the local level.

A key to the success of a two-tier national/local structure for protected landscapes is a two-way flow of information, experience and ideas between the two tiers so that the concept is not dominated by a top-down approach.

As will be seen in many of the case studies, it is often pressure from outside the communities which threaten the character of the landscape. Great care, therefore, needs to be taken to ensure that recognition of the values of the landscape at the national level does not lead to the breaking down of the traditional management structures under which the harmonious landscape evolved. In this respect, the goal should be to revive and/or reinforce those traditional structures or provide a structure which is equally effective with strong local leadership.

Cooperation and consultation

By their nature, protected landscapes involve implementation on a cooperative basis with local government and local people and effectiveness depends on the degree of support and partnership achieved with them.

The concept of partnership is built into the administrative structure for protected landscapes in most countries. However, the fostering of partnership needs to go beyond legislation, encouraging sympathy for the concept among all involved.

It is therefore important for local government to be involved to the greatest extent possible in implementing policies and in the process of consultation with local communities and interest groups. Through such an approach, local government should increasingly identify with the protected landscape concept and become more aware of the values placed on it by both the residents and the general public to the point of promoting the concept which, as has been mentioned, is the approach in France. The result should be enhanced local pride and greater consciousness of their stewardship role towards their area.

Cooperation and consultation, to be effective, must stem from a positive commitment to it. It should be real and not patronizing as, if it is seen purely as a procedural requirement of legislation and is grudgingly implemented, it will not be successful and effective implementation of the protected landscape concept may be jeopardized.

Where there are indigenous people with spiritual and historic links with the land, consultation procedures with them should be designed in ways which reflect their attitude to the land, are sensitive to that attitude and follow their cultural patterns. They should not be expected to conform to the general public procedures for consultation. This has been done with monastery-based discussions in the Himalayas of Nepal and, in New Zealand, on 'marae' according to the customs and protocol of the Maori people.

When there is a legislative requirement for public consultation, there should be a means of independent review of submissions arising from that consultation, perhaps by a small but representative group of members of the body responsible for the protected landscape.

As discussed earlier, some situations such as establishment of protected landscapes, adoption of management plans, or reviews warrant a two-stage process of public participation with an initial invitation for ideas including the release of issues papers. When submissions and views on these have been considered, a second round of public participation would follow based on the release of a draft proposal, plan or review.

The process of public participation should be seen as being as important as the end product. In this respect, Appendix 4 outlines an approach to the management planning process which can be modified and adapted for various processes of public participation.

The establishment process

A critical time for effective communication and consultation is the period leading up to the establishment of a protected area and this phase deserves particular comment.

While it is usual for legislation to lay down procedures for the establishment of protected landscapes, the legal requirements for public consultation should be regarded as the minimum and should be supplemented by a range of other mechanisms including those which are the normal practice in the region concerned.

Public meetings, meetings with local authorities and community and special interest groups should all be part of the process with any indigenous people involved being consulted in their traditional manner.

Because consultation is a vital part of achieving public support and acceptance and of gaining the maximum benefit from local and specialist knowledge born out of long experience in the area, the process should not be rushed and local initiative should be encouraged.

Box 8.2: LOCAL INITIATIVE

Experience with the US Nationwide Rivers Inventory shows the potential for local initiative. This Inventory was a listing of rivers potentially eligible for designation as Wild and Scenic Rivers. Although the inventory was compiled by technical experts following an administrative directive, there was also a nomination form which local residents could complete.

Subsequent to the Inventory, many communities nominated areas, some of which have subsequently been designated as part of the national system. The national list and criteria for the concept were important steps which then allowed local residents and communities to apply for recognition and designation. A similar approach could be used for Protected Landscapes.

Source: USNPS (1990).

Organizational and industry cooperation

There will also be a need for protected landscape bodies to maintain effective communication with interested regional and national organizations. Some of these may be generally supportive of protected landscapes such as conservation and recreation organizations, perhaps seeing themselves as watchdogs for the public interest. Some sympathetic organizations, such as National Trusts, may be significant landowners in the area and obviously a close partnership relationship with them is vital.

Some interests, such as property development and mining organizations, may see protected landscapes as a threat to development. Still others may have the potential for both positive and negative impact on protected landscapes. These include the farming, forestry and tourism industries. Ongoing liaison with all these industries is vital so that, as far as possible, negative aspects can be ameliorated and positive aspects encouraged.

It may be possible for protected landscape bodies to develop mutually agreed policies with such bodies. A good example of this is the adoption of agreed principles for tourism in protected landscapes developed and adopted jointly by the Countryside Commission and the English Tourist Board.

Box 8.3: PRINCIPLES FOR TOURISM IN PROTECTED LANDSCAPES

Tourism in protected landscapes needs to be guided and judged by all of these principles, if it is to meet tourist needs and protect the areas both now and in the future.

1. Conservation

The tourism industry can help to protect the distinctive landscapes and wildlife by supporting practical conservation measures. This can be achieved, for example, through joint initiatives involving the public, private and voluntary sectors.

2. Enjoyment

The activities and interests promoted by tourism should draw on the special character of the protected landscapes, with their many opportunities for

▶

quiet open air recreation and their distinctive beauty, culture, history and wildlife. Improved access for visitors should be sought where this is compatible with conservation requirements.

3. Rural economy
The social and economic well-being of the residents is an essential consideration in achieving the statutory objectives of protected landscapes, and employment in the tourist and related service industries is an important part of the economy of these areas. The tourism industry should support the economy of local communities through, for example, using employees, products and services from the locality and by supporting the skills and economic activities which are traditional to the protected landscapes.

4. Development
Appropriate facilities are needed to enable tourists to enjoy protected landscapes. All tourism development must respect the quality of the landscape and environment. Its scale, in particular, must always be appropriate to the setting. It should also recognize that some areas are valued for being wild and remote. Proposals for development should always be tempered by the capacity of the immediate site and surrounding landscape to absorb visitors. Development can assist the purposes of conservation and recreation by, for example, bringing sympathetic new uses to historic buildings and derelict sites and opening up new opportunities for quiet open air recreation.

5. Design
The scale, siting, planning, design, and management of new tourism developments should be in keeping with the landscape, and should seek to enhance it. The distinctive and highly valued character and landscapes of these areas will continue to evolve through small-scale changes. Major alterations to the landscape are unacceptable.

▶

6. Marketing
The tourism industry should use publicity, information and marketing opportunities to deepen people's enjoyment, appreciation, understanding and concern for protected landscapes.
Adapted from English Tourist Board and Countryside Commission (1990).

Staff/community relationships

Some participants in the International Symposium 1987 considered that protected landscape managers should have roots in the local community or, at least, identifiable links with it, and managers and staff should preferably have had practical experience in the locality rather than just have studied it from the outside.

Although this is not always possible, the point is valid that those involved in management are most effective if they can relate to the local people and are seen as part of the community and not apart from it.

9
Management of protected landscapes

The management of protected landscapes presents challenges which differ from those of most other categories of protected area.

Whereas the manager of a nature reserve or a Category II national park usually has direct management responsibility for areas which have little or no permanent human population, the manager of a protected landscape has a more indirect management role working with local government and often substantial resident populations.

Indeed, as protected landscapes may contain areas already protected for their natural, archaeological and historic values, the issues which arise in those types of protected area will demand the manager's attention along with the great variety of additional challenges of the protected landscape.

The nature of protected landscapes – the goal of harmonious interaction between people and nature and recognition that these inhabited landscapes are in delicate and dynamic equilibrium – underline the added dimensions of management.

Protected landscapes 'cannot be allowed to stagnate or fossilize. But change must be guided so that it does not destroy but will indeed increase the inherent values. This means for each protected area a clear definition of objectives, to which land use policies within it should conform. It also means a style of management that is sensitive to social and ecological conditions. This will be possible by building upon spiritual and emotional links to the land and by the operation of flexible systems of graded incentives and controls.'

Those who framed the Lake District Declaration added to this the comment that 'the protection of these landscapes depends upon maintaining within them a vigorous economy and social structure, and a population that is sympathetic to the objectives of conservation. It means working with people at all levels, and especially with those living and working in the area – the people most intimately affected by what happens to it.'

It is therefore vital for the manager to exercise his or her responsibilities sensitively and in cooperation with local communities and individuals.

Two-way communication with local people is vital to gain understanding of their concerns and aspirations and to convey to them the goals, benefits and obligations which derive from protected landscape policy and management. This is essential to enable the manager to operate in a climate of cooperation or, at least, acceptance because legislation or policies will not provide all the means of meeting protected landscape goals.

Consequently, willingness on the part of landowners and residents to cooperate in a voluntary manner to achieve policy goals is a key to success.

Broad community understanding and support can be far more effective in achieving cooperation than policing and penalty provisions in legislation. These may be necessary but they should be seen by the manager as very much the last resort.

Whatever legislative and administrative mechanisms are used, some common elements are important if effective management is to be achieved.

Box 9.1: REQUIREMENTS FOR EFFECTIVE MANAGEMENT

1. *A clear statement, preferably in legislation, of the purposes and objectives of protected landscapes which is then interpreted, made more specific, and applied to the particular situation of each area with public involvement.*
2. *Desirably, a national authority with the necessary expertise to develop policy and guide implementation and funding, working with skilled and experienced professional staff who provide advice and technical support to the authority and to the protected landscape management structure.*
3. *A similar structure at the local level with a representative body for the protected landscape serviced by professional staff, with a designated manager responsible for coordinating the effective management of the area concerned.*
4. *A means, appropriate to the situation, of achieving effective ongoing two-way communication between the protected landscape management structure and those who own property and/or live in the protected landscape, local government, recreational users and other interests.*
5. *A continuing review process at both the national and local levels to ensure that the purposes and objectives are being achieved, recognizing that the concept is dynamic rather than static and needs to have a capacity for policies and practices to respond to and adapt to new threats or take advantage of new opportunities and external social and economic change.*

Management principles

The approach to management of a protected landscape reflects the varied ownership of the land and the need to achieve the most effective degree of compatible and sustainable management for the range of values the area possesses.

Six general principles have been identified to guide management in endeavouring to achieve this.

Box 9.2: PRINCIPLES TO GUIDE MANAGEMENT

1. *Landscape protection is possible only when there is a vital and sound local economy with a positive perspective to the future. It is equally true that the distinctive landscape qualities of an area are themselves an essential element of the resources which can make sustainability possible. The management of a protected landscape is, in fact, the management of local economies and change.*
2. *Landscape protection is possible only with support from and involvement of the local residents. Therefore the concept of protection must be made attractive to local people, using a mixture of education, financial incentives and local powers of decision. Local people must see that protection provides positive advantages to them.*
3. *The basic ecological and cultural features of the landscape must be recorded, examined and protected.*
4. *In planning for development and management there should be available an adequate analysis of values, goals, impacts and options which can be put forward in non-technical terms for discussion with all concerned and which can then be used, in whatever form agreed, for ongoing decision-making.*
5. *The control tools used should be reasonably flexible and should respect the rights, needs and interests of local people.*
6. *There should be no illusion that a protected landscape can be managed as if it were an island in ecological, economic, political or cultural*

▶

> *terms. Its interests must be understood by and reconciled with those of the areas surrounding it.*
>
> *Source*: Summary proceedings, International Symposium on Protected Landscapes (1987).

Agriculture and forestry

Key resource uses which challenge management in most protected landscapes are agriculture and forestry. Both activities are major influences on the landscape and insensitive development can be destructive. The challenge for management is to seek to channel them in a manner supportive of the protected landscape.

Agricultural practices have often been pivotal in creating landscapes of secondary ecosystems which tended to be traditional and harmonious but which have experienced modification as a result of intensive mechanization and government policies designed to increase production and often to amalgamate small farm units.

Because agricultural impact on the landscape has historically been seen as positive, many States with protected landscapes have no statutory control over farming activities even though these can threaten the very values central to their designation. Examples include the drainage of wetlands, clearing of forests, removal of hedgerows, conversion of moorland into intensive grassland and use of modern materials for buildings, gates and fences instead of traditional ones.

Although national agricultural policy is outside the control of protected landscape managers, there is action the management can take. This includes education through displays, publications and talks and persuasion to cooperate in voluntary restraint, encouragement to convert to low impact tourism and craft ventures to compensate for lost farming revenue and, where funds are available, the use of grants or provision of materials and employment to modify farming practices and replace fences, gates, etc. with traditional materials. The use of farm conservation agreements or covenants formalizes these arrangements.

There is great value in interpretive and educational programmes being directed to local people as well as to visitors. In the Cevennes in France, field education activities are organized for local people as well as visitors and, in winter, the park management organizes social gatherings and meetings to explain and gain cooperation of local people. Reaching school children is important, too.

This type of approach is at the heart of success in protected landscape management.

A similar approach can be followed in relation to forested areas, particularly with those situations where only remnants of original forest cover remain. Cooperation with the owners of such forest land is essential to retain and, where possible, enhance the values of the forest remnants as essential components of the landscape and as refuges for biological diversity.

Mechanisms need to be developed which are designed to retain and enhance these values.

Box 9.3: GUIDELINES FOR MANAGING FORESTS ON PRIVATE LAND

1. *A survey and assessment should be made (at an early stage in planning) to identify and record the intrinsic value of forested areas for the conservation of flora, fauna and natural ecosystems.*
2. *In specific situations, where there are areas of outstanding and, possibly, unique value, high priority should be given to their long-term protection. This should be given preference over other forms of land-use and is particularly urgent in lowland forest. In such cases, owners should be encouraged to come to some form of arrangement which will ensure protection of the forest.*
3. *Before decisions are made to modify or transform untouched areas every consideration should be given to alternatives. This may include adapting already modified areas, for example, using unproductive gullies for woodlots. Alternatively, existing uses may be intensified or areas used for more than one purpose if these uses are compatible with protected landscape goals.*

▶

4. *Where the logging of indigenous forest is seen as acceptable, it should be done in a selective manner so as to*
   (a) maintain the soil and water conservation and wildlife values of the forest,
   (b) retain the forest structure,
   (c) allow for adequate regeneration, with timber extraction access carried out in ways that least damage soil, vegetation and watercourses and which minimize the visual impact.
5. *Owners should be encouraged to replant or promote the regeneration of forest species for those forests requiring rehabilitation, using species native to the area.*
6. *Owners should be encouraged to allow controlled recreation in their forests where the forest areas are large enough to absorb recreational use without any adverse impact.*
7. *Cooperation between the protected landscape agency and the landowners should lead to the preparation of plans or agreements for appropriate management of forest areas to ensure that their management is according to the best available principles of silviculture, and in such a way that the natural composition and structure are altered no further than is necessary.*
8. *Management should be monitored to assess whether the original objectives were reasonable and the management has been successful and the approach should be modified as appropriate.*

Source: Adapted from the Forestry Council (1980) and Poore and Sayer (1987).

Management planning

It is essential to have clear objectives and policies for management of each protected landscape and to have these available to guide all concerned to manage the area sensitively and to advance only those developments which are compatible. For each protected landscape, the management plan is the central mechanism to interpret and apply the legislation and general policy to the area.

It is usually a legislative obligation on those responsible for managing a protected landscape to prepare a management plan. Such a plan is not normally a formal part of the planning system of the country but takes account of that system and indicates in greater detail the approach to management within the protected landscape. The plan sets out the policies of the management body with such goals as:

1. to conserve the character and qualities of the protected landscape for present and future generations;
2. to have regard to the social, economic and cultural needs of the local communities; and
3. to provide for public use and enjoyment of the area.

The plan would include proposals for any land the management body owns and should set out the ways in which it aims to influence the management of other land.

The plan forms the basis for cooperation with other public and private landowners and with statutory and voluntary organizations with interests in the area to achieve conservation, community, and recreation goals.

Consultation in the preparation of the plan, along the lines indicated in Appendix 4, involves local people, local government and organizations in producing workable approaches and is essential to their cooperation. There should, of course, be opportunity for input by national as well as local non-government organizations and the legislation would normally provide for consultation with the national authority for protected landscapes (if there is one). Legislation may provide that the plan cannot be adopted finally without the approval or concurrence of the national authority. This would ensure national consistency along with local flexibility. In any case, a national authority and its support staff have much to offer in contributing out of their experience to the evolution of the plan.

Management plans and zoning provide the framework to guide management by giving it a systematic and consistent basis and communicating this to all concerned.

The plans need to take into account the results of research, not only in gathering information but in testing possibilities for innovation and they need to plan to meet and anticipate the nature of pressures for development.

Five particular types of pressure on protected landscapes have been identified.

1. Population growth with the consequent pressure for land for building and other development.
2. General national demands, such as for transportation, mining and military activities.
3. Agriculture, particularly in intensification of use.
4. Forestry, where this may be of a character alien to local tradition.
5. Recreation and tourism, especially in relation to major development of built facilities.

Because management of protected landscapes is not static and pressures change, plans are generally subject to regular review at, say, five yearly intervals to take into account changes in circumstances and ideas and any new legislation. Reviews should follow the same consultation procedures as the initial plan.

The management plan

A management plan would take the general policy matters discussed earlier and apply them to the protected landscape concerned.

A summary of the key elements of the 1988 review of the plan for the Peak District (UK) appears as Appendix 5. Peak Park holds a Council of Europe Diploma for its work in protected landscapes management. Its management plan reflects the role of the Peak Park Joint Planning Board with both a planning and, through its Parks Officer, a management role. The context of this plan can be appreciated from the case study section of this guide where Peak is one of the protected landscapes around the world discussed there.

The Pinelands National Reserve, in contrast, has a 'Comprehensive Management Plan' which has more of a planning than a management emphasis reflecting the fact that planning and guiding the management policies of others is the primary role of the Pinelands Commission.

Just as there is no single approach to the establishment of protected landscapes, so there is no single approach to management planning. For example, the French 'regional nature park' charters must, of necessity, be more descriptive than prescriptive in the absence of any planning powers vesting in the protected landscape authority.

However, all plans need to include:

1. the national and regional context of the area;
2. a description and inventory of the area sufficient to set the context for management;
3. the objectives;
4. the management considerations;
5. strategies to achieve conservation, recreation and rural development goals;
6. the means of implementation.

Desirably, plans should then seek to summarize the material in the manner most meaningful to those affected. If the protected landscape has a number of identifiable geographical sectors, then sector summaries will be very helpful in communicating its implications to those who live and work in those sectors.

As has been indicated, communication is a key to success in protected landscape and management. It is also a key to having a management plan with which people can identify and see as a springboard rather than as a straight jacket: the process of consultation throughout its preparation is thus most important and remains so in implementation and ongoing review.

Protected landscape management planning is not an end in itself but, like the protected landscape concept, it is a means of achieving more effective management of the countryside and its biological, cultural, social and economic resources. Goals can be identified and means of achieving them but, to be effective, committed people are needed to make the system work. The fact that there are many ways of doing this is clear and this is reinforced in the case studies which follow.

10
Benefits to residents of protected landscapes

Protected landscapes can bring a range of benefits and services to landowners and residents. These can range from the guarantee of an attractive living and working environment, assurance of the maintenance of a traditional or established way of life, the maintenance of services which might be lost with rural depopulation and economic benefits from tourism and recreation and associated economic opportunities. In some cases, there may be a range of financial incentives available to residents.

Economic and social benefits

The French approach illustrates the recognition of the many benefits – including economic and social – which residents may gain from the concept.

The fact that the initiative for establishment must come from the areas concerned demonstrates clearly that local people see distinct benefits from being part of an area which is identified nationally and internationally as one with special and distinctive qualities and which, therefore, merits special efforts to maintain the social fabric and economic viability of rural life.

The Normandie–Maine case study shows how local industries have been fostered in the production and marketing of cider and pears and in the restoration of forest crafts with up to half the residents in some villages employed in timber trades. Culture is kept alive through local museums, exhibitions and displays and there is help and advice in the establishment and operation of compatible tourism. The result is economic and social benefit through income generation and employment opportunities.

Box 10.1: PARTNERSHIP IN FRENCH PROTECTED LANDSCAPES

Managers of protected landscapes (regional nature parks) in France have no direct control over land use. Consequently, they must rely on a partnership approach, working with the local people with a strong emphasis on rural development and promoting local identity.

Techniques used in the Vercors 'protected landscape', a tilted limestone plateau south-west of Grenoble include:

▶

1. *fostering the establishment of non-profitmaking associations in fields ranging from compatible agricultural development to enhancing local communication;*
2. *developing training programmes for foresters, timber producers and craftspeople;*
3. *subsidizing a replacement farming service to allow farmers to attend training courses or take holidays;*
4. *cooperating with local groups in the promotion of conservation and local art;*
5. *coordinating development and promotion of tourist enterprises and establishing interpretive exhibits in cooperation with tourist attractions;*
6. *providing assistance to tourism and rural development in an alternative site to protect a proposed nature reserve from ski development;*
7. *fostering cooperative marketing of local products;*
8. *acting as a contact point with central government departments;*
9. *organizing cultural programmes and exhibits for residents and visitors.*

Source: Countryside Commission Information Bulletin (1985).

Advisory services in rural conservation

Inherent in the French example is the provision of advice and support for local residents and appropriate local industry.

Other countries, too, seek to influence farming practices and support rural activity by other than mandatory planning or fiscal controls. Frequently, this is attempted through provision of advice to farmers on conservation matters.

Thus, in the United Kingdom, the Ministry of Agriculture is under an obligation to provide farmers with advice on the conservation and enhancement of natural beauty and amenity of the countryside and to do so free of charge, a role complemented by the Countryside Commissions. In many tropical countries, a similar role is played by Rural Conservation Extension Services.

The desire of many living in economically productive and visually attractive landscapes to safeguard their properties from undesirable development can be facilitated in many ways, particularly under the protected landscape concept.

Landowners can work with the protected landscape agency, with government or local government organizations or with bodies such as national trusts as they seek to safeguard the future of the environment in which they live and work from outside pressures or from change should they have to relinquish ownership.

National Trusts, which may own land in their own right and, in fact, do in many protected landscapes, can also work with landowners to negotiate arrangements to protect conservation values on private land. Other examples involving agreements for the preservation of dry grasslands may be found in Sweden and some Swiss cantons.

Management agreements, covenants and easements

Management agreements may be taken for a fixed term or during the mutual agreement of the parties. However, in many cases there is no way they can bind third parties so that, if the land is sold, there is no guarantee the new owner will continue the terms of the agreement.

Greater security of protection is provided for in New Zealand legislation dating from 1977. This authorizes the government, local government and the Queen Elizabeth II National Trust to negotiate conservation or open space convenants with landowners. Under these, landowners may agree to commit themselves to preserve features of natural or historic or landscape amenity value on conditions and under management arrangements either written into the agreement or covered in an associated management plan. The incentive to the landowner lies in his or her sense of stewardship and, in some cases, covenants cover whole farms. These covenants have the strength that under the legislation they may be registered against the relevant land title and may bind successors in title. The Waipa District case study from New Zealand illustrates how the use of voluntary covenants has helped establish a *de facto* protected landscape. The existence of a protected landscape creates a situation where both the landscape body and the landowner may be motivated to maintain landscape qualities by such means as well as by easements which can provide a mutually useful landscape conservation tool.

Easements are rights obtained by the owner of one piece of land (the dominant tenement) over another (the servient tenement) which endure for the benefit of the dominant tenement and which bind all subsequent owners of the servient tenement. Public agencies owning land may therefore negotiate easements under which, for example, the owner of the servient tenement may commit himself or herself to manage the land in an environmentally sensitive manner, for example, by not felling trees, constructing buildings, etc. In the United States, landscape or scenic easements have been widely used as they can usually be acquired relatively cheaply as an alternative to outright acquisition of the land. The value of the easement for landscape protection varies, depending on the legal system of the country concerned.

A restrictive covenant is another device by which a landowner voluntarily accepts an obligation not to use land in a certain manner, that is, not to allow offensive uses of the land. In some circumstances, it may also bind successors in title to the land.

Under some legal systems, private agreements may be entered into with landowners by which the latter agree to refrain from various activities which would damage the landscape quality of their land. These may take the form of ordinary leases but, recently, the use of management agreements has become more widespread. These usually involve the participation of a public agency but are nonetheless private contracts. The obligations placed on the landowner (not to drain wetlands, fell trees, etc.) may be given in consideration, usually, of the payment of sums of money.

Financial incentives

The role of taxation as an incentive for conservation has already been discussed but other more direct fiscal incentives are being used to foster protection of features of importance in landscape conservation.

The South Australian Government's Save the Bush programme includes not only grants to private owners of bushland but provides an advisory service on the management of remnant native vegetation.

In 1989–90, the Australian Government announced a grants scheme to save native bush remnants with funding for survey and conservation projects for farming, educational, land care and environmental groups and local government.

Increasingly, nature and wildlife conservation programmes financed by bodies such as the World Wide Fund For Nature (WWF) have a rural conservation component to foster sustainable economic activity.

This is because financial and development incentives are increasingly being seen as important components of projects designed to conserve biological values in populated areas in tropical countries. International funding is becoming important in achieving harmonious interaction of people and nature in such areas where traditional sustainable approaches to hunting and gathering of forest resources are falling down, particularly under population pressures.

Box 10.2: THE CROSS RIVER PROJECT, NIGERIA

This WWF project, aimed to achieve sustainable land use in 400 000 hectares of forest in south-east Nigeria, recognizes that conservation policies cannot be implemented if they conflict with the economic needs of villagers. A core protected area has been identified with some 100 villages included in a 'support zone'.

Grants will be available to compensate for economic losses from protecting the core zone but subject to withdrawal of the grants from any village where residents do not respect the core area.

Traditional use zones will enable some villages to carry out controlled resource management for their exclusive benefit. As an added incentive, a corporation owned by villages in the support zone will be set up to capture profits derived from the conservation-compatible uses of the core and support zones. Local government and village liaison officers will have key involvement in the project.

Source: World Wide Fund For Nature (WWF) and Overseas Development Natural Resources Institute (ODNRI) (1989).

Financial incentives may involve large-scale projects, such as Cross River in Nigeria, or small-scale ones. An example of the latter is the arrangement entered into by a private conservation organization to finance the building of a school in the Western Samoan village of Falealupo in return for an agreement to maintain a communally owned area of rainforest with forest extraction limited to low key village use coupled with a village-operated eco-tourism project.

Thus, while protected landscapes are of national importance, they can also clearly bring a range of economic and social benefits to local residents.

Plate 1. *A view from Apple Pie Hill in the New Jersey Pinelands National Reserve. The Pinelands area comprises a large freshwater aquifer underlying forests, rivers, farms, and small towns. Courtesy of the New Jersey Pinelands Commission.*

Plate 2. *About 25 000 Maasai are resident within the Ngorongoro Conservation Area. They are primarily a pastoral people who graze an estimated 280 000 head of livestock over 75% of the conservation area. Courtesy of Jim Thorsell, IUCN, Gland, Switzerland.*

Plate 3. *Ghandruk village, an important centre in the Annapurna Conservation Area in Nepal, with the peak on Annapurna south and forested areas. Courtesy Annapurna Conservation Area Project.*

Plate 4. *Forests managed for thousands of years blend with cultural elements at Taishan, a landscape in the Shandong Province of China protected to recognize its significance as the site of one of China's most sacred mountains. Courtesy of P.H.C. Lucas, Chair, IUCN Commission on National Parks and Protected Area.*

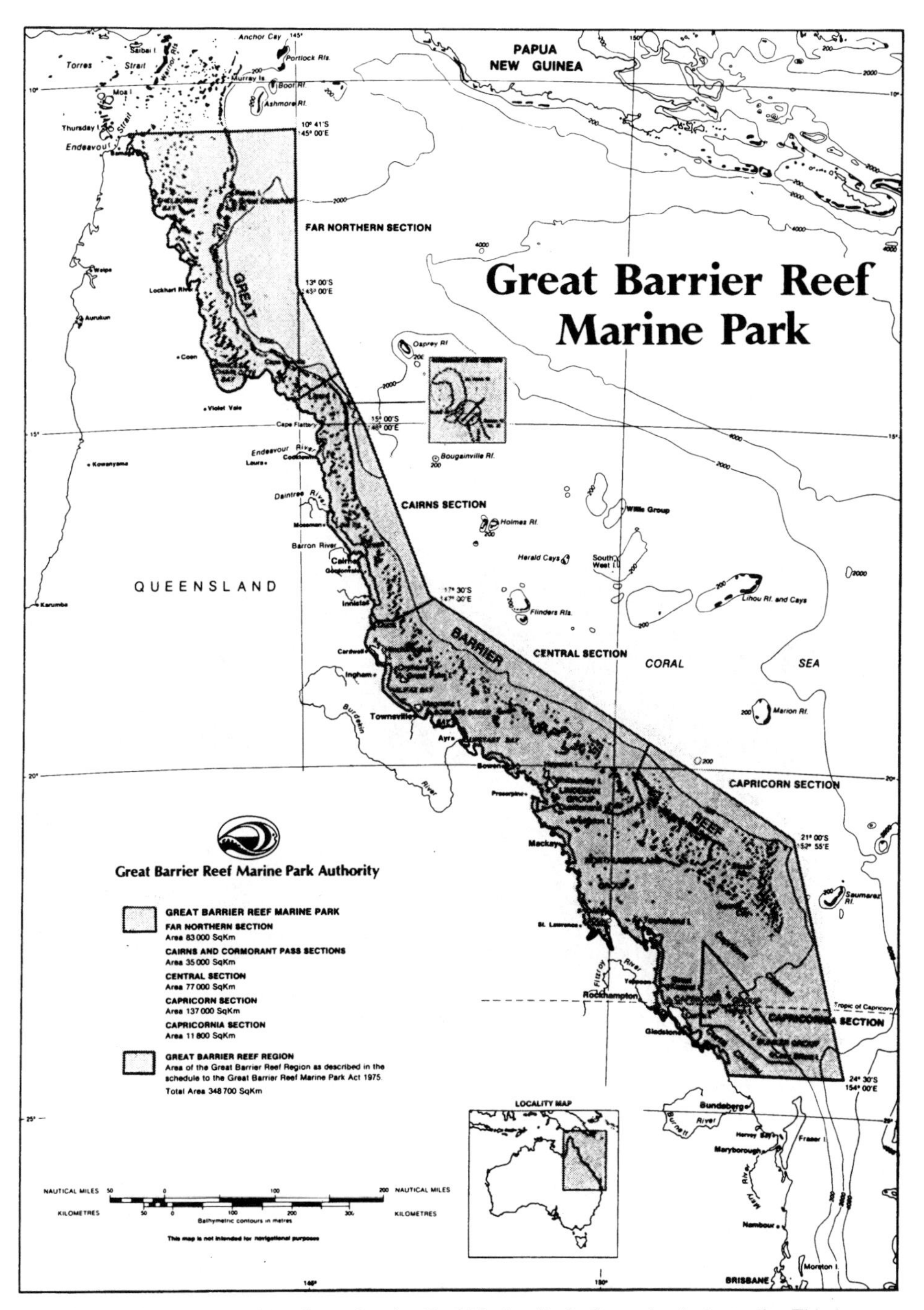

Plate 5. *Zoning plan of the Great Barrier Reef Marine Park, Queensland, Australia. This is a protected seascape covering 34 400 000 hectares and containing numerous coral reefs and islands. It is home to a great diversity of animal species, both marine and terrestrial as well as many plant species. Courtesy of the Great Barrier Reef Marine Park Authority, Townsville, Australia.*

Plate 6. *Combination of farmland and wetland in the Waipa countryside, North Island, New Zealand. The volcanic Mount Pirongia can be seen in the background. Courtesy of John Greenwood, Department of Conservation, Hamilton, New Zealand.*

Plate 7. *Waipa countryside showing natural forests and heavily modified vegetation in the form of grassland for grazing animals. Courtesy of John Greenwood, Department of Conservation, Hamilton, New Zealand.*

Plate 8. *Agricultural landscape around the village of Taddington in the Peak District National Park, Derbyshire. This was the first area in England and Wales to be designated a 'national park' on the Anglo–Welsh protected landscape model. Courtesy of the Countryside Commission, Cheltenham, UK.*

Plate 9. *Bielansko–Tyniecki Landscape Park in the valley of the Vistula River, southern Jurassic region, showing field patterns of agricultural holdings. Courtesy of Dr Z.J. Karpowicz, World Conservation Monitoring Centre, Cambridge, UK.*

Plate 10. *Pollution from industrial areas surrounding the Jurassic region in Poland. Courtesy of Dr Z.J. Karpowicz, World Conservation Monitoring Centre, Cambridge, UK.*

Plate 11. *The Ring of Gullion, Slieve Gullion, Northern Ireland. An area the Department of the Environment, NI, would like to declare an Area of Outstanding Beauty. This would aid the conservation of the area.*

11 The concept in action: case studies around the world

Introduction

The protected landscapes concept has been applied in different ways in many countries and the case studies which follow illustrate the variety of paths taken in a number of countries to reach a broadly similar goal – that of maintaining natural, cultural, social and economic values in living landscapes.

The case studies show how the principles outlined earlier in this guide have been applied in a variety of situations from the old world to the new.

The material for the case studies has come, in large measure, from the documentation of the Protected Areas Data Unit of the World Conservation Monitoring Centre based in Cambridge, UK, especially that prepared for the International Symposium 1987. It has been supplemented in many cases from material supplied by the management agencies themselves and, wherever possible, has been checked for accuracy and up-to-dateness with those agencies and with others.

The format used is designed to assess particular features of the concept illustrated by that specific case study. Each area is described briefly together with its particular values, how it came to be established and the legislative, administrative, planning and management mechanisms used. It is hoped that this will assist those interested in drawing from the experience of others and deciding how best the concept can be applied in their own situation.

It will be noted that whereas many of the case studies fit neatly into the IUCN Category V definition of 'Protected Landscape' and meet the principles outlined in the Lake District Declaration and in this guide, some do so only in part.

The Waipa County example from New Zealand lacks a nationally recognized identity and is based largely on voluntary mechanisms giving legal security, backed up by district planning. It is included because it shows that some protected landscape goals can be achieved in the absence of national recognition even though national legislation would give added status and security.

The Annapurna Conservation Area in Nepal is included as a carefully planned experiment in establishing a management regime for an area which, in effect, consists of a core Category II National Park and a large buffer area with considerable values in its own right qualifying as a Category V area.

Tanzania's Ngorongoro Conservation Area – classified as a Category VIII managed resource area – has many of the features of a protected landscape. Similarly, the Great Barrier Reef Area qualifies as a managed resource area but those who manage it also see it as a protected seascape.

Overall, the case studies show how the protected landscape concept contributes to the goals enunciated in the Lake District Declaration. In terms of the goals of IUCN, all the examples in the case studies make significant contributions, among other things, to nature conservation and to sustainability, often in situations which complement other categories of protected areas and demonstrate that people and nature can live and work in harmonious interaction with nature.

They, therefore, provide 'greenprints' for sustainable living for wider application both nationally and internationally.

It is hoped that these case studies and the many other protected landscapes around the world will inspire more countries to pursue the protected landscape pathway as one of the ways towards a better future for people and for nature.

Ngorongoro, Tanzania

Assessment: Ngorongoro Conservation Area (Plate 2) has important protected landscape elements accommodating as it does the Maasai people, their livestock and their evolving lifestyle along with a great variety and density of wildlife in a landscape with Ngorongoro Crater as a key feature. Initially part of the Serengeti National Park, it was excised from the park in 1959 and established as a separate 'conservation area' with an evolving management regime aimed at accommodating people and nature in one coherent, integrated, multipurpose protected area. The current needs are to integrate resource and wildlife conservation, rangeland and livestock management, tourism, and development for the benefit of the resident population. It is hoped that both an Ngorongoro Charter (issued at the highest appropriate level of Government) and a detailed management plan will be prepared in the near future.

IUCN Category: VIII; managed resource area.

Location: In the Arusha region of northern Tanzania, south-east of Serengeti National Park.

Physical characteristics: Includes the open plains of the eastern Serengeti rising from under 1500 m to the crater highlands of the volcanic massifs of Loolmalasin (3587 m) and Oldeani (3168 m). A key feature is the Ngorongoro Crater, one of the largest inactive unbroken calderas in the world which is not flooded. Includes Empakaai Crater and Olduvai Gorge. It is bordered in the west by Serengeti National Park and by the Rift Valley escarpment on the east. To the south lie densely populated agricultural lands and to the north the Loliondo Game Controlled Area. Rainfall varies from an average of 430 mm a year in some lowland areas to over 1700 mm a year in parts of the highlands.

Nature conservation values: Ngorongoro Conservation Area (NCA) is renowned for its great variety and density of wildlife with the Ngorongoro Crater its most spectacular attraction. A variable climate and diverse landforms and altitudes have resulted in several distinct habitats. Scrub heath and the remains of dense montane forests cover the steep slopes, the crater floor is mainly open grassy plains, and outside the crater are undulating grasslands. Among the large population of wild ungulates in the crater are wildebeest, zebra, eland, gazelles, black rhinoceros, hippopotamus and lion. Migrant species from Serengeti are numerous including over one million wildebeest which use the plains during the rains. There is a substantial bird population, including ostrich and the lesser flamingo, on the lake in Ngorongoro Crater and on Lake Ndutu.

Cultural values: The variety and richness of fossil remains has made NCA one of the major areas in the world for research on the evolution of the human species. Hominid footprints found at Laetoli are dated at 2.8 million–3.6 million years and the famous hominid fossil Zinjanthropus (now known as *Australopithecus boisei*) at 1.75 million years. Pastoralism, which continues with the Maasai people, has existed in the Area for over 2500 years.

Human population: Some 25 000 Maasai people reside in the area and they are increasingly becoming more sedentary and moving away from their traditional lifestyles. Nine villages are legally recognized within the Conservation Area. There are no residents in the Ngorongoro and Empaakai Craters or in the forest.

Economic activities: The Maasai residents graze some 280 000 head of livestock but cultivation was prohibited in 1975. The Ngorongoro Highlands are also important to surrounding pastoral communities, who use the area during times of drought. Tourism is important economically and has been increasing steadily since 1983. However, although the total number of visitors (both residents and non-residents) and the revenue earned from tourism are greater than at any point in the NCA's history, the number of non-resident tourists has not reached the levels recorded in the 1970s.

Establishment: Most of the area was included in Serengeti National Park when the park boundaries were finalized in 1948. Although the park was dedicated primarily to wildlife conservation, traditional cultivation and pastoralism were allowed to continue until cultivation was prohibited in 1954 sparking off a five-year controversy. In 1959, the area was divided into the present-day Serengeti National Park as exclusively a wildlife area with the balance becoming the Ngorongoro Conservation Area with pastoralism and cultivation permitted. However, cultivation was also prohibited here in 1975 because of concern at erosion and destruction of habitat for both wildlife and the domestic stock grazed by the Maasai people. NCA was inscribed on the World Heritage list in 1979 for both its natural and cultural values and was approved as part of the Serengeti–Ngorongoro Biosphere Reserve in 1981.

Area: 828 800 ha. Contiguous to Serengeti National Park (1 476 300 ha) and to Tanzania's Maswa Game Reserve and Maasai Mara National Reserve.

Land tenure: Government land.

Local administration: The Ngorongoro Conservation Area Authority was appointed in 1959 to assure multiple land-use was practised but differences developed between officials and the Maasai people. Apart from managing the area, the Authority provides services such as water supplies for pastoralists and livestock and purchases veterinary drugs and sells them at cost to residents. An *ad hoc* Ministerial Commission met over a 16 month period and reported in 1990 to the Minister for Lands, Natural Resources and Tourism on the adoption of a long-term management strategy for the area. During the consultation process, an IUCN workshop convened in 1989 suggested, among other things, closer integration between the Conservation Authority and local government (the Ngorongoro District Council), greater Authority involvement in community development, and the preparation of a management plan.

Land use issues: The Area's objectives were defined by ordinance in 1975 as:

1. conservation and development of natural resources;
2. promotion of tourism;
3. safeguarding and promoting the interests of the Maasai.

Conflicts between pastoralism and conservation have increased in recent years with three principal areas of conflict.

1. The Maasai are becoming more sedentary and their traditional lifestyle is changing with growing demands for modern housing, schools, clinics and other development, impacting on conservation and tourist values.
2. The human population is growing at a rapid rate, having risen from 8700 people in 1966 to some 25 000, placing increased demands on the Area's natural resources.
3. Livestock populations have not kept pace with the growth in the number of pastoralists to the point where the Maasai are no longer able to subsist on their traditional diet of milk and meat. Grain has become an important staple and there is growing pressure to allow cultivation and many illegal farms are appearing.

Various projects have been launched involving the Government of Tanzania, IUCN, and other conservation bodies and there is broad recognition of the importance of an integrated multipurpose policy to cover the variety of interrelated issues. The need for greater food security for residents has been identified as a matter of urgency and a long-term programme of sustainable use is seen as an essential prerequisite for conservation based on scientific information on ecological and social aspects. A major workshop on a conservation and development strategy for NCA was held in 1989, the Ministerial Commission on the Area's future reported to the Tanzanian Government in 1990 and the government response to the report is awaited.

Conservation management: A need is seen to improve scientific information on both ecological and social aspects of the NCA system through a continuing programme of research and monitoring. Other priorities are the control of poaching and protection of key species such as elephant and rhinoceros and conservation of the short-grass plains as critical habitat for the wildebeest migration. Total protection of forest, especially in critical water catchments and effective fire management are other important conservation management issues. Closer cooperation in management of the Area's natural and cultural values is needed. Hope is seen, too, for cooperation with surrounding communities in agro-forestry programmes to establish fuelwood plantations and enhance the supply of grain.

Human settlements: A need to stabilize the existing population in the present villages is seen with possible incentives to migrate out of the area to achieve a balance between population and resources. There is recognition that economic and social development of NCA residents needs to be undertaken with full community involvement. The goal would be to ensure improved community facilities and to develop economic potential through tourism, livestock production and other means. Scope is seen to ensure preferential employment of local people by the Conservation Authority and the tourist industry, the manufacture of craft products for sale and performances of traditional dances. The Maasai way of life is changing and it is seen as important for the Authority to work with them so that change is advantageous to the Maasai and compatible with Ngorongoro's natural and cultural values.

Recreation and visitor management: Visitors in 1989 numbered 107 000, of whom 47 000 were residents of Tanzania. This local interest is building up support for the Area's conservation. There are five tourist lodges on the crater rim and one at Ndutu. Vehicles and guides can be hired to go into the crater. The only interpretive centre – at Olduvai – is focused on the interpretation of the gorge and its excavations. Tourist fees contribute over 90% of the Conservation Authority's revenue.

Staffing and budget: The NCA Authority employs 319 people. Revenue in the 1988/89 financial year exceeded 194 000 000 Tanzanian shillings.

Management agency: Ngorongoro Conservation Authority, PO Box 776, Arusha, Tanzania.

Annapurna, Nepal

Assessment: The approach to management of the Annapurna Conservation Area (Plate 3) is innovative, relying on public participation and education and linking conservation with human development. It sets out to help villagers maintain or regain control over the natural resources on which they depend and with which they previously lived in harmony. Initial work is concentrated in the areas where most of the pressures are. While the aim is for the project to be self-supporting financially, establishment involves a significant boost from outside donations as a result of which the Annapurna Conservation Area has a larger budget than other protected areas in Nepal. On the other hand, it has a large and growing resident population and has significant elements of community and rural redevelopment in its management. It is a grand experiment, the success of which could have far reaching effects on rural lifestyle, landscape protection and conservation in and beyond Nepal.

IUCN category: The proposed Wilderness Zone is probably most appropriately Category II National Park while the balance could be regarded as Category V Protected Landscape.

Location: In the Annapurna Range of the Central Himalayas in central Nepal, north of Pokhara.

Physical characteristics: Dramatic alpine landscape with the catchments of three major river systems including the world's deepest valley (that of the Kali Gandaki at 1131 m) lying between two of the world's highest mountains – Dhaulagiri (8167 m) and Annapurna (8091 m). Climate varies from subtropical to alpine with over 3000 mm annual rainfall on the southern slopes of the Annapurna Himal and less than 300 mm in the rain shadow area such as Jomsom and even less in the Manang area.

Nature conservation values: Vegetation diversity is exceptionally high because of climate and altitudinal range varying from subtropical broad-leaved forests at lower altitudes through mixed broadleaved temperate rhododendron forests to coniferous forests on dry ridges with juniper species among those in subalpine and semi-desert areas. Fourteen endemic plants occur, the highest number from any protected area in Nepal, and over 100 varieties of orchids are recorded. Among 441 species of birds reported is Nepal's only endemic bird species, the spiny babbler (*Turdoides nepalensis*). Mammals include rare and endangered species such as snow leopard, musk deer, red panda and blue sheep.

Cultural values: Long-established human occupation by many diverse ethnic groups of Nepal, particularly Gurungs, Magars, Thakalis and Manangis. Many associated religious sites such as Muktinatha, Damodar Kund, Annapurna Deothali and Dudh Pokhari. Traditional lifestyles are maintained in isolated areas such as Upper Mustang and traditional agricultural and pastoral activity is practised elsewhere to varying degrees.

Human population: Some 40 000 people concentrated mainly on the southern fringe of the Conservation Area and along the valleys. More than 90% of residents are subsistence farmers who depend on forests for fuel, fodder for their livestock, and timber for construction. Servicing tourism is of growing importance.

Economic activity: The local economy is traditionally based on natural resources. Pressures have increased greatly consequent upon Annapurna's popularity for trekking with some 25 000 foreign tourists annually. Cultivation, grazing and forest extraction have increased with high demand for construction timber and fuelwood and new substandard villages have been established to service tourists. Livestock numbers have also increased with greater affluence, bringing problems of overgrazing. Government rural development programmes propose introduction of exotic grasses and livestock with risk of even greater modification.

Establishment: Establishment was initiated in 1985 by Royal directive from His Majesty King Birendra requiring the King Mahendra Trust for Nature Conservation (KMTNC) to investigate an appropriate protected status. A six month field survey supported by WWF/US produced a feasibility study which proposed the concept of a 'conservation area' designed to achieve conservation of natural and cultural values alongside harmonious development of tourism administered by a small unit, relying on local participation and self-sustained through entry and user fees. The concept was launched in 1986 by HRH Prince Gyanendra, chairman of KMTNC on the occasion of the 25th Anniversary of WWF. Management began as a project from 1986 ahead of legislation to define a conservation area as providing for 'the protection, improvement, and multiple use of natural resources according to principles that will ensure the highest sustainable benefit for present and future generations in terms of aesthetic, natural, cultural, scientific, social and economic values'.

In July 1989, the Nepalese Parliament passed a bill to amend the existing National Parks Act to authorize conservation area status as an addition to the types of protected areas already provided for. The KMTNC was entrusted with the responsibility to make a regulation for the management of the Annapurna Conservation Area.

Area: 260 000 hectares (for total project area).

Land tenure: Public land with many traditional and community rights overlain in populated areas.

Local administration: Local management is in the hands of a Conservation Area director appointed by KMTNC. The director works closely with local people and agencies involved in resource conservation and rural development. The overall programme for the area has five major components:

forestry/wildlife; alternative energy; community development; conservation education; and tourist information. As an example of implementation, a forest management committee has been established representative of the villages and has formulated rules for firewood collection based on traditional wisdom with technical advice, financial and legal support. Under voluntary agreements, villagers have bound themselves to a system of fines if their locally appointed guards find them cutting firewood outside the rules.

Land use issues: Key problems are overuse of forests because of pressures from population and tourism and overgrazing, particularly of higher-altitude pastures, as a consequence of overstocking. In addition, there is concern at substandard tourism development and construction of new villages lacking the character and environmental compatibility of the traditional ones.

To facilitate effective management, five zones have been identified.

1. Wilderness Zone which is a *de facto* IUCN Category II National Park covering all areas above the current upper elevation limits of seasonal grazing with a management prescription of no development and full protection.
2. Biotic/Anthropological Zone where modern influence or technology has not significantly interfered with or been absorbed by the traditional ways of life of the inhabitants. Visitor access is restricted to maintain the traditional way of life and landscape character.
3. Protected Forest/Seasonal Grazing Zone above the normal forest resource collection areas with sustainable low impact use.
4. Intensive Use Zone generally comprising the human settlement areas and characterized by intensive agriculture and forest use and including the main trekking corridor. Management policies include encouragement of traditional forest and pasture management systems with forest plantings to meet increased demands.
5. Special Management Zone comprising areas suffering from the environmental pressures of recent settlement and from tourism/trekking impact. Management policies are directed toward reversing present adverse environmental trends.

Conservation management: Nature conservation management is directed to ensure the permanent protection of viable examples of all the natural communities of flora and fauna with particular emphasis on endangered species and their habitats. To this end, extensive research is being undertaken with surveys of the habitat of blue sheep and snow leopard. Forest nurseries have been established, including one raising high altitude species, and projects are being implemented to lessen demand on natural resources. An aim is to produce trees for plantations for fodder and fuelwood and to assist forest regeneration.

Human settlements: The goal is to maintain or return to villages their traditional character while fostering community development through effective water supplies, sanitation, etc., and involving women in decision-making.

Recreation and visitor management: Approaches designed to lessen the impact of trekkers on the environment include establishment of a kerosene depot operated by a local management committee and solar heating in locally owned lodges to reduce pressure on fuelwood and to produce added income from hot showers for trekkers. To improve standards of accommodation, a Lodge Management Committee has been established which has fixed its own standards for services and prices for food and lodging with authority to penalize or dismantle lodges not meeting the basic standards. Education programmes are aimed at both lodge owners and visitors to gain their cooperation in resource conservation and increase understanding and respect for local traditions and religious sites.

Staffing and budget: The Conservation Area has four senior officers, 10 mid level assistants and 10 support staff with locally recruited staff under training for management. The Area has five field offices to cover 13 administrative units. The budget is US$250 000 a year. Entry charges of US$8 for international visitors are collected at the Immigration Office by project staff. These fees are designed to make the administration self-supporting ultimately.

Management agency: The Project Director, King Mahendra Trust for Nature Conservation, PO Box 3712 Babar Mahal, Kathmandu, Nepal.

Taishan, People's Republic of China

Assessment: The Taishan (Mount Tai) Scenic Beauty and Historic Interest Zone (Plate 4) is an unusual type of protected landscape in that there is no large resident population. However, the integration of an impressive landscape and thousands of years of human use has had a profound effect. The area's predominantly forest environment has been shaped by human impact in terms of both protection and reforestation. Conversely, it is clear that the Chinese perception of Taishan as a precious legacy stems from its mix of natural and cultural values producing an harmonious interaction of people and nature.

IUCN category: V; protected landscape.

Location: In central Shandong Province just north of Tai'an City.

Physical characteristics: Mt Taishan rises abruptly from the vast plain of central Shandong. The geologically ancient mountain forms an impressive rugged, rocky and wooded backdrop to the city of Tai'an. The protected area extends north-west along the valley floor. The altitude ranges from about 150 m above sea level to Jade Emperor Peak at 1545 m. Average annual rainfall at the summit is 1132 mm, about one and half times that at the foot.

Nature conservation values: Vegetation covers 80% of the area and the flora is diverse with 989 species recorded. Broadleaf trees and conifers predominate and there are many varieties of fruit trees. Some very old and famous trees date back many centuries. Animal species recorded number over 200 with 122 species of birds and a rare red-scaled fish is found.

Cultural values: Cultural values are extremely high and cultural relics include memorial objects, ancient architectural complexes, stone inscriptions and archaeological sites. There are 22 temples, 97 ruins, 819 stone tablets and 1018 cliffside and stone inscriptions. Taishan is one of the birthplaces of

Chinese civilization, evidence of human activity dating back 400 000 years to Yiyuan Man of the Palaeolithic Period. By Neolithic times it had become a significant cultural centre and for over 3000 years Chinese emperors of various dynasties made pilgrimages to Taishan for sacrificial and ceremonial purposes. Rock inscriptions bear testimony to these visits as do ancient trees, planted by emperors as long ago as 2100 years. Many old trees are named, numbered and respected and have their histories recorded in archives. Renowned scholars including Confucius composed poetry and prose and left their calligraphy on the mountain. More recently, Mao Zedung left his calligraphy inscribed in stone. One of China's most sacred mountains, Taishan is a focus for Buddhist and Taoist pilgrimages and watching the sunrise from Taishan is a goal of many Chinese.

Human population: A town and several villages are located in the valley floor section, their residents being mainly occupied in farming.

Economic activities: The major economic activity in the protected landscape is tourism with Taishan attracting some 2 million Chinese visitors and pilgrims a year with some 13 000 foreign visitors. Historically, agriculture was important and is still practised on the valley floor. Much of the area previously cleared for agriculture has been reforested, generally using local species.

Establishment: Because of its cultural and religious significance, Taishan has been protected to some degree throughout history, except in times of war and revolution.

Protection measures culminated in Taishan's establishment as a Scenic Beauty and Historic Interest Zone in 1982 by the State Council of the People's Republic of China. Criteria for such zones include a pleasing environment, considerable size and tourist, cultural or scientific interest.

Legal protection is afforded to the natural and historic values under the Republic's Cultural Relics Protection Law, Forest Protection Law, regulations concerning the administration of Scenic Beauty and Historic Interest Zones and various local regulations and administrative decrees.

Taishan was inscribed as a World Heritage site (Cultural and Natural) in 1987.

Area: Approximately 50 000 hectares including buffer zone.

Land tenure: The State owns the land which, in terms of the Constitution of the People's Republic of China, means 'the whole people'.

Local administration: Local administration is the responsibility of the Administration Committee of the Scenic Beauty and Historic Interest Zone of Mt Taishan. This is a coordinating body drawing together various elements of government at the level of city administration. It has a planning, coordinating and management function and is responsible for all construction within the area. It operates with various sections handling administration, labour and capital, tourism, planning, landscape, health, industrial and commercial management, safety, construction and finance together with an archive, museum and display function.

The local administration operates within the oversight of the central Ministry of Urban and Rural Reconstruction and Environment Protection and within regulations promulgated by the State Council of the People's Republic.

Land use policies: China-wide policies laid down by regulation for Scenic and Historic Interest Zones include establishment of buffer zones, active measures to maintain wildlife habitat and protect forests with ancient and famous trees totally protected and management of other areas of forest subject to consent of the administering body. Developments for tourism and other purposes must be in harmony with the environment and new structures are not permitted on sites of high value.

Regulations for management of the Mt Taishan area promulgated in 1985 identified a core area known as the first-class preservation zone and a buffer under the administration of the Taishan Forest Farm. The Administration Committee has power to order the dismantling of unauthorized buildings or buildings that pollute the environment and to order renovation of buildings that are not harmonious with their surroundings. Natural, cultural and historic features are protected. Quarrying is forbidden in the first-class preservation zone.

Conservation management: Apart from famous, historic and rare trees being protected, regulations provide for controls on felling or thinning of trees, collection of firewood, medicinal herbs and other forest by-products. Collection of flowers and plants is not permitted and hunting is forbidden. Visitors are excluded from an area where research is being undertaken into the rare red-scaled fish by Shandong Agricultural University.

Human settlements: The town and villages must comply with the regulations and, where there is a need for natural resources such as rock supplies, sources are being developed outside the protected area.

Recreation and visitor management: The numbers of visitors and their concentration on the long-established stone steps up the mountain to the sunrise-viewing site and its associated temples, accommodation and visitor services cause major problems of over-use and a concentration of often incongruous structures and stalls. Local pressures and the absence of an effective management plan saw the construction of a cable-car system to a point near the summit but generally sited away from the view of those walking up the mountain. Plans are now in hand to remove some of the more incongruous buildings and bring improved control on commercial activities associated with tourism on the mountain.

Staffing and budget: Not separately identified.

Management agency: The Administration Committee, Mt Taishan Scenic Beauty and Historic Interest Zone, Tai'an City, Shandong Province, People's Republic of China.

Fuji–Hakone–Izu, Japan

Assessment: This protected landscape, established legally as the Fuji–Hakone–Izu National Park, combines public land and privately owned land subject to special planning controls stemming from the central government level. With Mt Fuji as the focus, visitor pressure with its associated tourism infrastructure is a major problem. A potential problem is airfield development on the Izu Islands. They demonstrate issues which stem from the national level and show that recreation, if not appropriately constrained, and major and incompatible development projects, can prejudice the protected landscape, even if there are mechanisms to achieve a balance between natural and cultural values and the economic activities of local residents.

IUCN category: V; protected landscape

Location: On south central Honshu, about 100 km southwest of Tokyo: in four parts Mt Fuji, Mt Hakone, the Izu Coast and peak on Izu Peninsula, and Izu Island chain.

Physical characteristics: Fuji (3776 m) is Japan's highest mountain and its volcanic cone rises from the plains. Mt Hakone to the southeast is a highly eroded extinct volcano while both the Izu Peninsula and the chain of islands which extend from it are volcanic in origin with some continuing activity. Coral reefs occur off some islands. There are areas of virgin forest throughout with alpine vegetation above the tree line on Mt Fuji and over 500 species of vascular plants recorded on Miyake Island, 21 species or varieties being endemic to the Izu area. Mammal fauna is largely palaearctic including the Japanese macaque, various bats, wild pig and Honshu sika and the bird fauna is very diverse. The Izu Islands have the Japanese woodpigeon, a scarce species with very limited range, and the southern islands are the only known breeding area for Izu Island thrush and Ijima's willow warbler. Altitude extends from sea level to 3776 m. The climate varies from subtropical warm temperate to alpine, and annual precipitation is approximately 2000–3000 mm.

Cultural values: Fuji is one of the three most sacred mountains of Japan with a shrine built at its summit in 1604 the centre of the Fuji-kou cult and the object of pilgrimages. There are other temples and shrines in the park. It is said that Mt Fuji represented the highest altitude achieved by humans until the rise of alpinism in the European Alps in the 19th century.

Human population: The park has a 'fairly high resident human population', the areas of highest density being on the perimeter. All the main islands in the Izu group are inhabited.

Economic activities: Economy is based on tourism, forestry, agriculture and livestock rearing with fishing an important industry on the coast and the Izu Islands.

Establishment: 1936 under National Parks Law for protection of places of natural beauty but primarily in response to recreational and visitor pressures. Additions made in 1938 and 1955 with the Izu Islands added in 1964 under the amended Natural Parks Law of 1957.

Area: 122 686 ha of which 7728 ha is designated a special protection zone and first class special zone.

Land tenure: Approximately 48 500 ha is private land; the balance is state and public land.

Local administration: Local administration works under guidance of the central Environment Agency and its Nature Conservation Bureau which, in turn, has five divisions of planning and coordination, natural parks planning, conservation and management, recreational facilities, and wildlife protection. National parks are designated landscape areas of national importance – 'areas of the greatest scenic beauty' – and include natural and man-influenced environments in multiple ownership with all land controlled by park law provisions and planning restrictions. Administration focuses restriction on actions which may harm the landscape and on provision of facilities for public use. Regulations for protection drawn up for each park and signed by the Director General of the Environment Agency zone areas to restrict activities harmful to the landscape. They include 'ordinary zones' (buffer zones) which restrict mining and changes of water level but permit forestry and fisheries and have few restrictions on commercial/industrial activities including tourism. Greatest protection is in 'special protection zones' kept free of development and including protection for particularly significant natural or cultural sites. 'Special zones' allow some development within a priority for protecting landscape (Classes I to III). At the local level, the park has a national park office and several ranger offices with nine park rangers in total. Local government has greater involvement in protected landscapes of regional importance, quasi-national parks, and those that are representative or important locally, prefectural national parks.

Conservation management: The park is important for its wildlife and has been the subject of a diversity of research studies. The major conservation problems come from the great visitor pressure especially on the Mt Fuji section of the park. The Izu Islands are also under threat through reforestation and a 1986 plan to build a military airport on Miyake-jima with potential threat of disturbance to breeding birds and soil run-off killing fringe coral reefs. Access to the summit of Fuji is officially open to the public only during July/August to reduce visitor pressure and reduce risk of accidents. An average of 400 000 people walk to the summit each year.

Human settlements: The density of local population adds greatly to pressures from tourism but the zoning system contains the spread of settlements and associated commerce and industry.

Recreation and visitor management: Records of over 20 million visitors make it the most visited area of its kind and there is an enormous range of facilities including hotels, inns, camp and picnic grounds, mountain and nature trails, cable cars, visitor centres, natural history and historic craft museums, a botanic garden of native plants, aquariums, golf courses, ice rinks, etc. Some of the larger islands are popular deep-sea fishing and holiday resorts with Miyake-jima of the smaller islands accessible to visitors attracted by its birdlife.

Staffing and budget: Most funding comes from central Government through the Environment Agency to cover staffing and subsidy for public facilities such as trails, visitor centres, parking, camp sites, toilets, etc. Some funds for facilities and equipment come from prefectural governments.

Management agency: Fuji–Hakone–Izu National Park, Headquarters Office, Environment Agency, Kojiri, Hakone Town, Kanagawa Prefecture, Japan.

Great Barrier Reef, Australia

Assessment: In 1975, Australia established a system of management for the Great Barrier Reef region (Plate 5) which embodies the objectives of the World Conservation Strategy and which can be regarded as a Protected Seascape. It provides a structure for environmental protection, planning, coordination and integrated management of a vast area of islands, reefs, shoals and sea with significant commercial fishing and navigation as well as tourist and recreational use. The system is seen to be working because there is a public commitment to have the area protected as a priceless national asset; there is over-riding legislation which prevails over conflicting provisions of other legislation; there is an ongoing programme of public education and participation in decision making; all uses must be taken into account in the management system; and the policy is to provide for sustainable use of resources in a manner which does not jeopardize long-term conservation of the Reef.

The regime set in place is dependent on cooperation between the Commonwealth and Queensland Governments. The latter has jurisdiction over most islands and waters above low water mark, whereas the Commonwealth is responsible for all waters, reefs and shoals below low water mark. The Great Barrier Reef Marine Park Authority is the principal adviser to the Commonwealth government on the care and development of the Marine Park. Effective cooperation with Queensland is assured through the structure of the Authority, a joint Ministerial Council and a consultative Committee. Maintenance of this cooperative relationship is vital. Fortunately, for the Reef, its protection is a policy of all the major political parties within Australia, and therefore no Government can afford to leave itself open to criticism by not lending its full support to the Reef's protection.

One of the 'on the water' problems associated with the two-Government system is this question of jurisdiction. Both the Authority and the Queensland agencies have recognized the potential for confusion to the user public in parallel and complementary zoning provisions. As a means of reducing this confusion, the agencies have agreed to the development of a joint interpretive map series depicting the zoning provisions of both plans together with relevant and appropriate written material deemed necessary for users.

A further aid to integration is the fact that the Queensland National Parks and Wildlife Service is responsible for day-to-day management of both the Great Barrier Reef Marine Park (GBRMP) and the Queensland marine parks and island protected areas under Queensland State jurisdiction.

IUCN category: VIII; multiple-use management area.

Location: The Reef extends 2300 km down the east coast of Australia from Papua New Guinea in the north to south of the Tropic of Capricorn, although the GBRMP stops in the vicinity of Cape York.

Physical characteristics: The GBRMP includes 2900 individual reefs, including 760 fringing reefs, which range in size from less than one hectare to more than 10 000 hectares. There are some 300 cays, 213 of which are unvegetated, and some 250 continental islands. The form and structure of the individual reefs which make up the Great Barrier Reef show great variety. Two main classes of reefs can be defined; platform or patch reefs resulting from radial growth, and wall reefs resulting from elongated growth, often in areas with strong water currents. There are also many fringing reefs where the reef growth is established on subtidal rock of the mainland coast or continental islands.

Variations in physical characteristics with latitude allow the area to be divided into three distinct sectors.

1. The northern sector characterized by shallow water covering the Continental Shelf (generally less than 36 m) with an almost continuous line of elongate wall reefs on the outer edge of the Shelf. Between this line and the mainland is a great variety of patch reefs, many with cays including low wooded islands which are coral cays and carry mangrove communities.
2. The central sector, with the depth of water ranging from 36 to 55 m is characterized by scattered platform reefs and fringing reefs on the mainland coast and coastal continental islands.
3. The southern sector is the deepest part of the Shelf, with depths ranging

> to 145 m and has a tightly packed maze of wall-like reefs in the north and, to the south, where the tidal range is less extreme, the outer Shelf is characterized by small, tightly packed reefs. In the extreme south is a group of large patch reefs, many of which have vegetated coral cays.

The climate is tropical, rainfall varies greatly from year to year and in geographical distribution, and tropical cyclones occur irregularly.

Nature conservation values: As the largest system of corals and associated life-forms anywhere in the world, it is an area of enormous faunal, floral and geomorphological diversity. There are believed to be some 400 species of coral including branching, plate-like, encrusting and mushroom corals. Other invertebrates include over 4000 species of molluscs. There are some 1500 species of fish, six species of turtle including sites of world importance for green and loggerhead turtles, cetaceans include humpback, minke and killer whales. Dolphins abound and there are many dugong. The cays and continental islands support 242 species of birds with 21 breeding colonies of sea birds and 109 breeding sites of land birds.

Cultural values: The Great Barrier Reef and, in particular, the northern sector, is important in the history and culture of the Aboriginal groups of the coastal areas of north-east Australia and there are large, important Aboriginal or Torres Strait Islander archaeological sites on a number of islands. The Reef figures prominently in European exploration from 1770 when James Cook's *Endeavour* ran aground on a reef subsequently named Endeavour Reef. Like the Cook expedition, many voyages were for research and survey.

Human population: Minimal but intensive human use is made of the Reef.

Economic activities: Commercial fishing and tourism, recreational pursuits (including fishing, diving and camping), aboriginal fishing, scientific research, and shipping all occur within the Area. Tourism is the largest commercial activity in economic terms. The number of people who visit the Great Barrier Reef annually was estimated at just over half a million in 1986. These people together spend over 2.2 million visitor days a year on the reefs and islands of the Reef and commercial boating expenditure alone is estimated at A$250 million annually.

The reef area supports a number of significant recreational and commercial fisheries. These fisheries which extend along the full length of the Reef include otter trawling for prawns, shovel-nosed lobsters (known as 'bugs') and scallops; trolling, gill and drift netting for mackerel and other pelagic species; hand-lining for reef fish; gill netting for coastal pelagic fishes such as trevally (mackerel), queenfish and threadfin salmon; mudcrabbing; collection of aquarium fishes, coral, beche-de-mer (edible sea cucumber), and trochus (mother of pearl); and trolling for big and small game species, such as marlin, sailfish and tuna.

People living in Aboriginal communities in the reef area have access to the marine and near-shore resources which played an important role in the Aboriginal economy during the past several thousand years.

Establishment: Numbers of small islands in the reef area were already terrestrial protected areas when serious conflict on and about the Reef and its management arose in the 1960s when the people of Australia became aware of, and objected to, proposals to drill for oil and to mine limestone on the Reef. The ensuing controversy revealed that the Reef was treasured by many Australians for its uniqueness, biological diversity, beauty and grandeur.

In 1973 the Commonwealth Parliament passed the Seas and Submerged Lands Act which established overtly Commonwealth jurisdiction over, and title to, the seabed below low water mark outside State internal waters. This Act was challenged by some of the States but its constitutional validity was upheld by the High Court in 1975.

In the same year, the Commonwealth Parliament passed the Great Barrier Reef Marine Park Act to provide for the establishment, control, care and development of a marine park through provisions which include:

1. establishment of the Great Barrier Reef Marine Park Authority (GBRMPA) consisting of three members, the Chair and one other are nominated by the Commonwealth Government, while the other member is nominated by the Queensland Government;
2. specification of the Authority's functions including recommending areas for inclusion in the Marine Park, carrying out research, preparing zoning

plans, establishment of education and management plans, and anything incidental to these functions. It also gives the Authority the power to perform its functions in cooperation with Queensland and its agencies;
3. establishment of a Consultative Committee made up of representatives of government, industry and community bodies with statutory provision that at least one-third of its members are nominated by the Queensland Government, one member by the Authority and the remainder by the Commonwealth Government. In practice, the Commonwealth and Queensland Governments have equal numbers of nominees on the Committee;
4. prohibition of drilling and mining in the Marine Park except for approved research purposes;
5. giving the Act, its regulations and zoning plans primacy over conflicting provisions of both Commonwealth and Queensland legislation, except in relation to the navigation of ships and aircraft.

A GBR Ministerial Council has also been established to coordinate policies of the Commonwealth and Queensland Governments in the Reef Region.

Since 1975, 98.5% of the Great Barrier Reef region has been declared a Marine Park and in 1981 the Reef Region (including the areas of state jurisdiction) was inscribed on the World Heritage list as a natural site.

Area: 34 400 000 hectares.

Tenure: Exclusive rights to explore and exploit the seabed of the Continental Shelf beyond the territorial sea are vested in the Commonwealth subject to certain limited rights conferred on third parties. Seabed inside the outer limits of the 3-mile territorial sea is vested in the State of Queensland, subject to a number of reservations including the operation of the Marine Park Act. Within the State of Queensland, title is vested in the State apart from public lands owned by the Commonwealth. Some land is held by private persons.

Local administration: Local management of both the Marine Park areas and the island protected areas is handled by staff of the Queensland National Parks and Wildlife Service, directly for the Queensland Government in relation to State-controlled protected areas and as agent for the Marine Park

Authority for the Marine Park areas. For the latter, the Service works to plans and policies drawn up under the special policy and administrative structure developed for the Marine Park. Additional surveillance and enforcement are provided by Coastwatch (the national aerial surveillance operators) and the Queensland Boating and Fisheries Patrol.

Resource use issues: There is some conflict between the various users of the Reef and those who wish to see it maintained in a near pristine state forever. Some uses of parts of the reef have already reached levels that appear to exploit fully the productive capacity of the system, e.g. bottom trawling for prawns. Run-off from islands and the mainland containing suspended solids, herbicides, pesticides, nutrients, and other materials is also a major concern. However, it is not yet known if this represents a real threat to the Reef.

Concern to prevent unacceptable ecological impact is paramount in the Authority's management of tourism development. The types of biophysical environmental impacts which may be associated with reef based tourism operations include discharge of waste, litter and fuel, physical damage to reefs from anchors, people snorkelling, diving and reef walking, disturbance of fauna (especially seabirds), over-fishing or collecting. All of these may be managed to some extent by design, prohibition or limitation.

Conservation management: The Marine Park Authority's primary goal is 'to provide for the protection, wise use, understanding and enjoyment of the Great Barrier Reef in perpetuity . . .'. Consequently, the approach is that any use of the area should not threaten the Reef's essential ecological characteristics and processes.

The Authority's policy is to maximize opportunity for human enjoyment by keeping regulation of activity at the minimum considered necessary to achieve conservation objectives. The Park is divided into sections and these are divided into zones with different protection status. Typically, highly protected zones are adjacent to moderately protected, or buffered zones which separate them from zones available for a broad range of uses.

Zoning plans within the Marine Park are a form of broad-scale management plan and the planning process leading to the application of a zoning plan within the Marine Park involves a number of discrete steps which are outlined in the Act. The major feature of the zoning process is a public participation programme. Divided into two stages, the programme first invites public input early in the planning process to help construct a draft zoning plan and later invites comments on the suitability of the draft plan produced. Public participation assists the Authority in gaining information on the resources and uses of the Park, identifying management issues and separating conflicting uses. The programme allows interested members of the public to have a say in the development of a zoning plan and encourages Marine Park users to take some responsibility for the management of the Park. Public participation is vital as day-to-day management of such a huge area depends to a large degree on public cooperation.

Zoning plans under the Act have now been prepared for all four sections of the Marine Park. These are the Mackay/Capricorn, Central, Cairns and Far Northern Sections. It is the policy of the Authority to review zoning plans after almost five years of operation to take account of changing circumstances. The most liberal zone is the general use 'A' zone covering all permitted activities; general use 'B' zone excludes trawling; Marine National Park 'A' zone excludes trawling, commercial netting, collecting and spearfishing; Marine National Park 'B' zone adds to the exclusions bait netting and gathering, crabbing and oyster gathering, line fishing and traditional hunting, fishing and gathering. Only research is permitted in the Scientific Research Zone and, in the Preservation Zone, only research which cannot be carried out elsewhere in the GBRMP.

The second major management tool available to the Authority additional to zoning plans is the power to issue or refuse permits for a broad spectrum of activities. These include tourist facilities, education and research, aircraft operations, discharge of waste, collecting, installation and operation of moorings and traditional hunting and fishing. The main purposes of the permit system are to encourage responsible behaviour in users, separate potentially conflicting uses, gather data for management and place reasonable limits on particular operations consistent with sustainable use. Appropriate conditions under which a permitted activity may be conducted are attached to each permit and are arrived at through a permit assessment procedure developed by the Authority.

Environmental impact statements – or Public Environment Reports – may be required for proposals with significant impacts. Public advertising may also be required. Permit assessment fees are charged for tourism permits.

The Authority is currently introducing a third management planning tool, that of Reef Use Plans, which are intended to provide specific guidance for permit issue at individual reefs. Development of these plans also involves a period of public participation.

The GBRMPA funds research relevant to the Marine Park. Research is undertaken on contract by other agencies for the Authority and covers a range of subjects relevant to the Marine Park.

Recreation and visitor management: Tourism is allowed to occur under permit within all except preservation and scientific research zones, that is, in 99.8% of the Marine Park. While the amount of the Marine Park designated free from tourism or fishing may seem low, it must be recognized that the Marine Park encompasses large areas of open water, so the proportion of reef areas so designated is, in practical terms, much higher. Resort guests make extensive use of reefs and waters for recreational activities, including fishing, diving, and snorkelling, water sports, sightseeing, reef-walking, and some shell collecting. Large, stable, high-speed catamarans provide day trips to islands and outer reefs. The education of reef users to appreciate the reef environment and exercise care when visiting the reef is a major element in managing and protecting the resource. The Authority produces materials for public education and assists tourist operators in developing programmes that are conservationally and educationally focused for visitors. The Authority also operates a major Aquarium in Townsville which serves, among other things, as a facility to educate the public about the Reef. A tourist attraction in itself, the Aquarium attracts about 300 000 visitors annually. The Aquarium development is an example of how innovative management can contribute to the enjoyment of a natural feature and to the development of a sustainable tourist industry. While this facility provides a 'reef experience' for visitors who might not otherwise be able to visit the Reef, the Aquarium is also aimed at encouraging people to visit the Reef itself. The cost of running the Aquarium is met from entry fees, sales in the Aquarium shop and donations.

Staffing and budget: Staff of the Authority are Commonwealth public servants and in 1989/90 numbered 101, 24 of whom were employed in the self-funding Aquarium. Because staff resources of both the Authority and the Queensland agencies involved are stretched to their limit, management resources concentrate in areas of intensive use. The Queensland National Parks and Wildlife Service faces similar pressures and currently has 92 staff involved in the day-to-day management of the Area.

In 1989–90 the Commonwealth Parliament appropriated A$9 266 000 for the Authority's programmes other than the Great Barrier Reef Aquarium. In addition, A$1 million was appropriated for the Aquarium as an advance to provide working capital. Additional funds amounting to A$946 850 were carried forward from the 1988–89 financial year and applied towards 1989–90 activities. Under the cost-sharing arrangements for the day-to-day management of the Marine Park, an amount of A$2 191 140 was received from the Queensland Government during the year. Receipts from other sources including interest, permit assessment fees, contribution for baseline and monitoring studies and the sale of education materials amounted to A$942 694 and earnings of A$1 618 829 by the Aquarium.

Programme expenditure for the year, other than expenditure on the Aquarium, amounted to A$12 162 034 or 91% of funds available.

Management agencies: Great Barrier Reef Marine Park Authority, GPO Box 791, Canberra ACT, Australia 2601.
Queensland National Parks and Wildlife Service, PO Box 190, Brisbane North Quay, Queensland, Australia 4000.

Waipa, New Zealand

Assessment: Waipa (Plates 6 and 7) demonstrates how far along the road to a protected landscape a sympathetic local authority, operating cooperatively with national conservation bodies can go towards achieving landscape protection. However, the lack of any national status or backing of the protected landscape by central government legislation leaves some elements vulnerable in periodic district scheme reviews. This situation was acknowledged in a general discussion paper released in 1988 by the Minister of Conservation who referred to 'a growing world recognition of the concept of protected landscapes' and suggested that, with diffuse provisions for protection of cultural landscapes in existing New Zealand legislation, 'it could be appropriate to include in conservation legislation explicit provision for protecting cultural landscapes on private, Maori or public lands'. Meantime, remnant forest has been saved from logging, some has been protected in perpetuity, other areas have interim protection and there are more positive attitudes to the protection of forest in the landscape although this cannot be said for attitudes to wetland conservation.

IUCN category: Waipa does not qualify but is cited as an example of an approach towards achieving protected landscape goals by voluntary legal protection coupled with statutory planning mechanisms.

Location: In the Waikato district of the North Island of New Zealand, centred on the town of Te Awamutu and just south of Hamilton city.

Physical characteristics: A distinctive and attractive landscape of mature pastoral character with low rolling hills, flat alluvial land, peat bogs and lakes and, on its boundaries three mountains – the volcanic remains of Kakepuku, Maungatautari and Pirongia. Remnant forest areas and peat lakes in the farmed areas are dramatic features in the landscape, often seen against a backdrop of the forested volcanic hills.

Nature conservation values: Waipa District's original vegetation has been heavily modified in the development of grassland farming. Consequently significant indigenous forests on the volcanic slopes along with the remnants of forest and other indigenous vegetation among the farmland have great conservation value. The mountain forests are predominantly a mixture of podocarps and broadleaved trees while many of the forest remnants are dominated by the podocarp kahikatea. There are peat bogs surrounded by scrub which provide excellent bird habitat for several secretive and scarcer species. A wide variety of more common indigenous and introduced bird species is found in the varied habitats.

Cultural values: The Waipa District has been settled for over 1000 years with Maori settlers from Polynesia establishing cultivations and village (pa) sites. European land settlement came in the 1800s and in the 1860s Waipa was the scene of land wars. There are well-preserved Maori pa sites, some with fortifications and ancient cultivation areas from pre-European times. There are also numerous battle sites and remains of fortification from the Waikato campaigns of 1864.

Human population: The population of the district is 36 000. Predominantly rural with concentrations in several small village communities and with some closer rural residential settlement in the vicinity of Hamilton city.

Economic activities: Predominantly a dairy farming area with about 85% of the land under cultivation. A trend towards diversification has brought an increase in the area in crops, orchards and vineyards. The area could reasonably be described as one of the most intensively farmed districts in New Zealand.

Establishment: While no legally constituted protected landscape has been established, progress towards protection of landscape values has evolved since the late 1970s. Then, the county engineer, concerned about loss of peat lakes convened a meeting of government conservation and university personnel who were able to convince the then Waipa County Council to consider working to protect remnant stands of indigenous forest as well as the wetlands. The council engaged a team with skills in protected areas, wildlife conservation, botany and limnology to undertake field work which was translated into the listing of 'conservation areas' on private land in the district planning scheme. In the resultant landowner response to the scheme, the Council consulted the Queen Elizabeth II National Trust, a body set up by legislation to promote the voluntary protection of open space and landscape values in New Zealand. The Trust developed an education programme publishing booklets on the Waipa landscape and its values and the county council distributed these to all residents to raise awareness and encourage landowners to enter into open space covenants with the Trust to protect and provide for effective management of those values. All landowners with areas on their properties considered deserving of protection were visited on the Council's initiative and were later invited to a joint Council/Trust seminar to explain protection mechanisms. The voluntary open space covenants with no compensation payment involved are entered into in perpetuity by landowners with the Trust and are registered against land titles binding successors in title. They are backed by a district planning scheme aimed to support landscape protection.

Area: 112 900 hectares.

Land tenure: Mainly privately owned. Protected areas in public ownership include parts of Manungatautari and Pirongia mountains and some of the peat bog and lake areas.

Local administration: Local administration is handled through the normal operations of the now Waipa District Council and particularly through its district planning scheme which has legal force under central government planning legislation. The Council employs a Reserves Officer whose efforts are backed up by central government Department of Conservation staff.

Land-use policies: The general aim of the district planning scheme is to promote a well-balanced and rational pattern of compatible land uses which allow for a wide choice of lifestyles and activities while protecting community interests. The scheme includes policies for resource conservation, environmental quality and preserving and recognizing Maori cultural values. The environmental quality aim is to protect and enhance the special qualities and variety of the district's landscape. Specific provisions in the scheme call for preparation and implementation of management plans for key environmental resources such as lakes and for council consent to any 'substantial alteration' to natural features. This covers disturbance of regenerating and mature forest areas, bulk removal of topsoil, any work likely to destroy the dominance of any feature lending character to the area, any work affecting any natural water course, river, harbour or lake, and any significant change in land contours.

Conservation management: Additional to land-use policies, there have been cooperative schemes for fencing of forest remnants to keep livestock out and encourage regeneration which was also fostered by a council nursery which supplied native plants to revegetate damaged forest remnants. The council has also offered conservation incentives of exemption of rates (land taxes to local government) on covenanted areas and assisted fencing forest areas through job creation schemes. The National Trust has negotiated permanent covenants with 19 landowners covering 105 hectares to protect forest remnants, lakes and other wetlands with further covenants under action. The publicly owned protected areas are managed directly by the Department of Conservation or local government with the Council extending its protected areas by purchasing and fencing 105 ha of forest adjoining Maungatautari.

Human settlements: The district planning scheme aims to promote urban consolidation by avoiding unnecessary expansion of existing settlement and preventing sporadic urban development in rural areas.

Recreation and visitor management: Apart from normal community recreation facilities, the emphasis for visitors is on recreation based on walking tracks in Pirongia Forest Park and Maungatautari Mountain Scenic Reserve and on driving for pleasure. The district possesses some of the most visually attractive pastoral landscapes in New Zealand and these can best be viewed from roads which follow ridges. To protect these views, the district scheme identified areas of restricted development needing special consideration for siting of buildings, etc. with the aim of enhancing the ridge roads as view corridors.

Staffing and budget: See *Local Administration.*

Management agency: Waipa District Council, Private Bag, Te Awamutu, New Zealand.
Associated with:
Queen Elizabeth II National Trust, PO Box 3341, Wellington, New Zealand.
Waikato Regional Office, Department of Conservation, Private Bag 3072, Hamilton, New Zealand.

The Pinelands, United States of America

Assessment: A sound scientific basis, a flexible approach and the partnership between federal, state and local units have proved fundamental to The Pinelands success. These factors, together with excellent public communication and appropriate back-up legal powers have meant that decisions have withstood opposition and many legal challenges and are increasingly gaining broader acceptance. An economic analysis of the Comprehensive Management Plans' impact on the region has shown that neither the economic vitality of the Pinelands nor the fiscal integrity of its municipalities has been hindered while development has been channelled into less environmentally sensitive areas.

IUCN category: V; protected landscape.

Location: New Jersey, extending from the hinterland of 11 major drainage basins to the Mid-Atlantic Coast and Delaware Bay.

Physical characteristics: The Pinelands or 'pine barrens' (Plate 1) are a gently rolling landscape with extensive largely unbroken forests of pine, oak and cedar. They are a patchwork of forests, rivers, farms, crossroads, hamlets and small towns underlain by a large freshwater aquifer which feeds slow-moving streams and the marshes and bays of southern New Jersey. With its sandy soils, the area is very sensitive to pollution. The altitude extends from sea level to 60 m. Rainfall averages between 107 cm and 117 cm and winter temperatures average 0–2° C; summer 22–24° C.

Nature conservation values: Contains extensive surface and ground water resources of high quality with 59 species of amphibians and reptiles and 91 species of fish. Habitat types include salt marsh, white cedar swamp, sphagnum bogs, cranberry bogs, upland pine–oak, pygmy pine plains and hardwood swamp. The flora has over 800 species of vascular plants of which 580 are native, 270 introduced, five endemic and 71 endangered, threatened or undetermined. In addition, 34 species of mammals and 299 species of birds are recorded.

Cultural value: Indians hunted and fished the area 10 000 years ago. Early European settlers extracted 'bog iron' for cannon shot for the Revolutionary War, timber provided raw materials for boat building, charcoal and other industries. The sand and gravel deposits attracted the china and glass industries, some elements of which survive, and one of the old abandoned towns (Batsto Village) has been restored. Traditional row crop farming existed on the edges of the forest and cranberry and blueberry cultivation became major industries.

Human population: 495 000 year round residents in all or parts of 52 municipalities and parts of seven counties.

Economic activities: Agriculture, including blueberries and cranberries as well as row and field crops, is extremely important. Other major industries are recreation, resource extraction, construction (mostly on the periphery), shellfishing and public service.

Establishment: Growth pressures in the 1970s threatening the environment and patterns of 300 years of human use coincided with the State of New Jersey's interest in protecting the landscape and the identification of the Pinelands as an area of national concern and significance by the Department of the Interior. Federal legislation in the National Parks and Recreation Act 1978 established the Pinelands National Reserve 'to protect, preserve and enhance the significant values of the land and water resources of the Pinelands area'. The legislation was designed to encourage and assist the State of New Jersey and its units of local government to develop and implement a 'comprehensive management plan ... to assure orderly public and private development' consistent with the purposes of the reserve and to encourage coordination of government programmes affecting the land and water resources of the Pinelands area. A management structure was provided for and, in 1979, the New Jersey legislature supplemented the federal law with its Pinelands Protection Act establishing the Pinelands Commission confirming a moratorium on incompatible development during the planning process and requiring mandatory local compliance with the proposed comprehensive management plan which was to be completed within 18 months.

Accepted as a Biosphere Reserve in 1983.

Area: 438 210 hectares of which, under the State legislation, 148 928 hectares is identified as a core 'preservation area' to be most stringently protected from the impact of future development with a surrounding 'protection area' with a mix of environmental features, farmland, hamlets and towns where development is allowed in a manner which would not degrade the 'essential character' of the Pinelands environment. Since 1979, some 27 000 hectares of critical habitat has been purchased.

Land tenure: Two-thirds is privately owned and one-third public lands mostly in the Preservation Area. State-owned areas include parks and forests and Federal properties include wildlife refuges and military installations.

Local administration: Administration is by a three-level partnership involving federal, state and local governments coordinated by a 15 member Pinelands Commission as an independent state agency. The Commission's structure set in the Federal legislation includes one member appointed by the Federal Secretary of the Interior, one member from each of the seven counties in the reserve appointed by the respective counties and another seven members appointed by the Governor of New Jersey. The Federal law also provides for the Commission to include residents of the reserve who represent economic activities of the area such as agriculture and residents of New Jersey who represent conservation interests. These provisions are reinforced in the New Jersey State legislation which also established a Pinelands Municipal Council representative of each municipality in the Pinelands area to act in an advisory capacity.

The Commission exercises complete land-use jurisdiction in the Pinelands and no state facilities may be developed in the Pinelands without the Commission's review and approval. The Commission also coordinates with other state planning agencies to ensure compatible land use on the National Reserve boundaries.

The Federal Department of the Interior must approve amendments to the Comprehensive Management Plan and municipal governments must submit local master plans and zoning ordinances for Commission approval so that all use of public and private land must ultimately be reviewed by the Pinelands Commission.

The Commission's three main concerns have been:

1. to protect resources under its jurisdiction;

2. to create scientifically and legally defensible regulatory provisions;
3. to draft policies that would permit flexible implementation and be responsive to legitimate local and individual need.

The Commission is essentially a planning and not a management agency and relies on the existing federal, state and local government organizations and, where appropriate, the private sector to undertake management of conservation, recreation, etc.

Land use policies: The Comprehensive Management Plan (CMP), developed with extensive public participation, includes a natural resource assessment; a land use capability map and a comprehensive statement of land use management policies; a study of appropriate public uses of land; a financial analysis; a programme to ensure local government and public participation in land use decisions; and a programme to put the plan into effect.

Land use management areas established under the CMP are:

1. Preservation Area District ('The Heart of the Pines') for resource-related uses such as cranberry and blueberry agriculture, forestry, recreation and fish and wildlife management with limited residential development.
2. Forest Areas, environmentally sensitive lands that display many qualities similar to the preservation area.
3. Agricultural Production Areas, larger concentrations of conventional agricultural lands with related commercial and residential activities.
4. Special Agricultural Production Areas largely limited to cranberry and blueberry farming or native horticulture.
5. Rural Development Areas, already semideveloped and planned for moderate density development compatible with the Pinelands environment.
6. Regional Growth Areas adjacent to already developed portions of the Pinelands where uses may be determined by municipalities to achieve an assigned growth density.
7. Pinelands Towns, traditional communities primarily outside regional growth areas where municipalities may determine future land uses compatible with the existing character.
8. Pinelands Villages, settlements with cultural and historic ties to the

Pinelands where municipalities may designate land uses compatible with the existing character.
9. Military and Federation Installation Areas with current uses recognized.

Flexibility has been the key to implementation of the CMP. Each section of the plan is preceded by 'flexibility language'. This allows management area limits to be moved if municipalities can convince the Commission this is appropriate while a provision for 'letters of interpretation' means that anyone can ask the Commission how the plan applies to an unusual circumstance or a use not anticipated by the plan. This allows the Commission to define its intent and apply the plan to unique situations rather than being tied to the precise language of regulatory sections.

Conservation management: The scientific basis, seen as critical to planning and management, was developed in the process of plan preparation and is built on the continuing work of the Rutgers University Division of Pinelands Research and an acquisition programme was commenced for critical habitat areas, mainly in the Preservation Area. Public education in this and other fields is both a statutory requirement and an important management tool.

Human settlements: Most local government plans now conform to the CMP and the character of existing settlements is being maintained by guiding major development to identified 'regional growth areas'. This is facilitated by the transferable development (TDR) programme under which Pinelands Development Credits (PDCs) were allocated to lands in the Preservation Area and to prime agricultural lands and these PDCs were transferable to any of the designated regional growth areas.

Recreational and visitor management: Recreation opportunities are varied, including exploration of old abandoned towns, hiking, boating and canoeing the inland waters and swimming and fishing in both inland waters and the Atlantic Ocean. Camping is very popular. The state park system manages a major part of the considerable recreational demand.

Staffing and budget: An Executive Director heads a staff of 51 (two-thirds professional, one-third support). The annual operating budget is approximately US$2.5 million. Initially, the Federal government provided US$26 million for planning and acquisitions (of which US$25 million has been used for acquisitions from Federal Land and Water Conservation Funds) with US$33 million from State 'Green Acres' fund. An additional authorization for US$14 million for acquisitions has been approved by the US Congress.

Management agency: The Pinelands Commission, State of New Jersey, PO Box 7, New Lisbon, New Jersey 08064, USA.

Martinique, Martinique

Assessment: Le Parc Naturel Régional de la Martinique represents an attempt to transfer the protected landscape concept from Europe to the Caribbean with similar goals of protecting the natural environment, promoting sound resource management and rural development, and the promotion of cultural values. The park administration is also a channel for a wide range of environmental concerns about Martinique from broad issues such as shaping public opinion to support conservation action to specifics in the protection of watershed values. This role must be balanced against the primary task of seeking to guide management of the area in a way which optimizes nature protection and promotion of cultural values with economic and recreational benefits while operating within the limits of management resources and of a tropical island environment.

Achievements are mostly in the areas of environmental education, recreation, economic development (mostly in relation to the tourist industry and handicrafts), and as a framework for interagency cooperation, especially with respect to land-use planning. Less has been achieved in conservation of flora and fauna, in scientific research, or in protection of endangered species within the framework of the park. The lack of binding authority and a limited budget are seen as problems in this respect.

These problems do not make the concept invalid for island situations. Rather they illustrate the need for different approaches to meet different situations.

IUCN category: V; protected landscape.

Location: On the lesser Antillean island of Martinique in the Caribbean. In three separate parts from the north-west tip (City of Grand'Rivière) to the town of Fort de France; in the south from Baie de Fort de France to Pointe des Salines; and on the east coast, Caravelle Peninsula.

Physical characteristics: The area includes the volcanic mountainous and hilly region of the island with the active volcano, Mount Pelée, as well as coastal cliffs, beaches and reefs but excludes the cultivated lowlands. Altitude ranges from sea level to 1397 m at Mount Pelée. Average annual rainfall varies from 7620 mm in the Martinique highlands to 500 mm. Lies within the hurricane belt.

Nature conservation values: There is diverse vegetation ranging from dry woodland and mangrove in coastal areas, cactus scrub in low rainfall areas to rain forests from 500 m and cloud forests above 800 m. Mahogany has been introduced into the forests at different altitudes. Green turtle and hawksbill turtle breed along the southern shores and the mangrove habitats are rich in molluscs and crabs. Frigate birds, sooty terns and brown noddies are among the coastal avifauna and the Martinique oriole is characteristic of the humid forest zone. The *Ramphocinclus brachyurus* is endemic to Martinique.

Cultural values: There is some evidence of prehistoric Arawak Indians but today's inhabitants are of French, African and Carib origin. Historic ruins include 17th century distilleries and colonial houses and the ruined city, St Pelée, is on the slopes of the volcano.

Human population: Some 80 000 people live in the protected landscape; 360 000 on the island.

Economic activities: The economy is based on tourism, aquaculture, fishing, and the banana and rum industries with beef cattle bred and grazed on the grass pastures in the south.

Establishment: Martinique's status since 1946 as an Overseas Department to France means that the legislation of metropolitan France applies. The area was established as a parc régional naturel in 1976.

Area: 70 150 hectares.

Land tenure: The montane areas are under private ownership as is most of the coastal strip. The balance of the area is under public ownership including 517 ha in La Caravelle réserve naturelle and 250 ha in the wetland reserve of Fort de France.

Local administration: A board coordinates management in an advisory role. It consists of representatives of the municipalities, local administrations and communes. The Caravelle Peninsula reserve is protected by a local wardening system while other major natural habitats are managed by the Office National des Forêts. Many initiatives are undertaken in nature and cultural conservation in conjunction with other organizations and publications are used as an educational tool.

Land use policies: The area is managed primarily to safeguard the natural and cultural heritage of the area including the maintenance and development of the rural economy. Areas are zoned to safeguard the different land uses. Support is given to rural development through helping the establishment of small rural enterprises.

Conservation management: The area is managed to conserve wildlife and hunting is prohibited on the wetland of Baie de Fort de France, an area considered to be of international significance for migratory birds as the ornithological reserve of Ilets de Ste Anne. Sewage pollution is a problem in Baie de Fort de France and inland erosion has led to excessive siltation in the coastal areas. The shallow lagoon areas are excessively fished by local people and *Tibouchina chamaecistus* (V) is being depleted through picking of its flowers in the high altitude regions of the park. Extensive surveys have been undertaken in the marine environment.

Human settlements: The area includes 34 communes.

Recreation and visitor management: Activities include 33 hiking trails, swimming, scuba diving, snorkelling, and spear fishing. Golf, cycling and sailing are also catered for. An ecomuseum for the park is housed in the Ancien Collège Agricole in Tivoli and there are nine museums on the island covering themes such as geology, conchology, ornithology and the rum industry. Recreation and interpretation programmes are targeted first to local residents and are also seen as important for visitors.

Staffing and budget: The administrative base is Fort de France but staff are also located at several protected natural and cultural areas within the overall protected landscape.

Management agency: Parc Naturel Régional de la Martinique, Ancien Collège Agricole, Tivoli, BP 437, Fort de France 97200, Martinique.

Peak, United Kingdom

Assessment: Peak is the classic type of protected landscape in a developed country with sophisticated planning systems. Much of the success of its management is the product of many years of fostering a sense of partnership by enhancing communication and relationships with landowners and residents and with park users (Plate 8). With the substantial surrounding population, recreation pressures remain a major challenge and innovative approaches are being used. Through very diverse programmes, the Peak Park Joint Planning Board is showing how three major purposes – protecting landscape, providing for visitors, and serving the well-being of local residents – can be pursued together and in mutual support.

IUCN category: V; protected landscape.

Location: North Midlands of England at southern tip of Pennine Range. Ringed by large industrial cities including Sheffield, Manchester, Derby and Stoke-on-Trent.

Physical characteristics: The area comprises varied countryside of rocky crags rising to moorlands of heather, with great ridges ('edges') above softer landscape of open and wooded limestone valleys ('dales'). Altitude ranges from 104 m to 636 m above sea level and rainfall from 890 mm to 1524 mm.

Nature conservation values: Natural flora and fauna and geological and geomorphological features are an integral part of the Peak district's values. Important 'seminatural' wildlife habitats of self-sown native species exist in ancient woodlands and old permanent grasslands. Moorlands are influenced by and dependent on low-intensity agricultural activity (usually sheep grazing and grouse rearing). Many species of birds are found and among the mammals is the mountain hare, found on peat moors and reintroduced for sport after disappearing from the area around 400 BC.

Cultural values: Stone Age hunters inhabited the area as evidenced by prehistoric 'henges' or stone circles such as Arbor Low. A Bronze Age landscape is found at Big Moor and several Iron Age hilltop forts and 'Celtic' field systems remain. Romans came to seek lead and left their mark in straight roman roads. After the Norman Conquest, hunting preserves were established and Peveril Castle above Castleton was built as a Norman stronghold by the son of William the Conqueror. Attractive villages are built of locally quarried limestone and gritstone and the historic dual economies of lead mining and sheep grazing have much influence on the landscape. For example, lead mining has left small irregularities in the plateau grasslands where miners dug the veins of lead ('rakes') and around some lead mining villages there are stone-walled enclosure fields built in the 18th and 19th centuries.

Human population: 38 000 of whom 45% are in service industries (including tourism, 15%), 28% in manufacturing and 10% in agriculture (1981).

Economic activities: These comprise sheep farming along with dairy and beef, with moors favoured for grouse shooting. Farmers are diversifying into farm holidays using farmhouses, converted barns and caravan and camping sites typifying the growing importance of recreation and tourism. Limestone quarrying continues as does other relatively low impact mineral extraction. Coniferous woodlands planted by the Forestry Commission and others cover a significant area, especially around reservoirs in the north and east.

Establishment: As the first national park in England and Wales, the Peak was a product of lengthy consideration of the concept first examined by the UK Government in 1929. Post World War II reconstruction saw a report to the government in 1945 by John Dower on how the park concept might apply in England and Wales. His definition of 'an extensive area of beautiful and relatively wild country' being managed for the nation's benefit to protect characteristic landscape, provide appropriate recreation, protect wildlife and cultural values and maintain established farming use was adopted. It was applied later by a Government agency, the National Parks Commission, set up in 1949, which identified areas which met the criteria. The Peak was the first established.

It was designated in 1951 and awarded the European Diploma of Council of Europe in 1966. The generally oval-shape protected landscape excludes areas already marred by industrialism, mainly limestone quarrying. It includes one national nature reserve and 54 sites of special scientific interest. There are many scheduled archaeological sites, 28 conservation areas in villages and towns and over 2100 listed buildings.

Area: 140 382 hectares.

Land tenure: Most land is in private ownership. The National Trust owns 10%, the Peak Park Board 4% and water authorities 1.5%.

Local administration: The Peak Park Joint Planning Board, which sets policies and serves as the statutory planning authority, has two-thirds local authority representatives and one-third appointed by central government. A National Park Officer serves the board and directs management of land the Board owns and the services it provides such as information and study centres and associated publications, information and interpretive services, hostels, caravan and camping sites and cycle hire. A ranger service assists in liaison with visitors, owners, tenants and agencies operating in the area;

assists the voluntary search and rescue services; carries out practical countryside work; and enforces bylaws.

The park administration influences use and management of other land, working through statutory planning procedures, programmes for rural development, conservation and environmentally sensitive tourism, cooperative management agreements, etc. Key documents guiding policies are the Structure Plan guiding land-use and the National Park Plan which sets out management policies to conserve the area's character and qualities, to provide for public enjoyment and have regard for local community need.

(NOTE: The local administrative structure is the most direct of the protected landscapes in England and Wales. Peak has a Joint Planning Board with members nominated by local authorities or appointed by the Secretary of State for the Environment: it is independent of the local authority structure and has its own staff headed by the National Park Officer. Lake District has a similar 'independent' joint planning board but is staffed partly by Cumbria County Council. For the remaining parks, much servicing is carried out by local government and the national park committee is legally a committee of local government. In each case there is close liaison with the Countryside Commission exercising its broader responsibilities. For example, the Commission provides advice on policy and on financial coordination and must be consulted on National Park Plans and on the appointment of National Park Officers.)

Land use issues: The Board's approach to landscape and nature conservation is guided by its division of the park into three conservation policy zones incorporating the park's various landscape elements. The Natural Zone where controls are tightest includes gritstone moorland, shale-grit seminatural moorland, limestone hills and limestone dales. The general policy for the Natural Zone emphasizing wild seminatural conditions is applied in ways appropriate for each of these landscape elements. The Rural Zone, where policy is for development and management of agriculture and forestry to respect and enhance traditional characteristics, includes appropriate specific policies for trees, woods and forests and for enclosed farmland. The third zone consists of the human settlements.

The strongest land-use conflict is with mineral extraction with more than 50 active quarries, mines or other sites. The park administration has recognized the national and regional interest in purer limestone and rarer minerals not available elsewhere; and proposals for their extraction have generally been accepted subject to restoration measures. However, proposals for mineral workings to meet the market for aggregate material have usually been resisted because adequate supplies are available elsewhere.

Changes in agricultural policies in the European Community are resulting in a shift from an emphasis on incentives for increased production to a joint goal of environmental conservation and production. Experiments have begun, within an Integrated Rural Development scheme, to show up the mutual support between social, economic and environmental aspects and to gain landowner and resident support. Grants encourage new rural enterprises linked to agriculture such as drystone walling, woodland management, food and timber processing, local marketing and conversion of redundant buildings to appropriate visitor use.

The park administration maintains close liaison with forestry interests to ensure use mainly of native species in plantings in designated 'natural zones'. On its own 6000 hectares of estates, the board seeks to provide a model of multipurpose management.

Conservation management: Regular liaison is maintained with the Nature Conservancy Council to clarify roles and coordinate nature conservation action generally, and specifically on nature reserves and sites of special scientific interest. A basic habitat survey gives priority to rare or most threatened habitats and planning is in hand to set up a single coordinated biological records system. Close cooperation is also maintained with private conservation organizations, some of which own land which they manage for wildlife values. Information is given to landowners on conservation values and the Board has entered into management agreements with many farmers to safeguard a variety of sites including flower-rich meadows, wetlands and woodlands.

Human settlements: The 150 villages in the Park are mainly built in local stone. Each has its distinctive character which policy aims to maintain. Villages include many 'Conservation Areas' under general planning legislation as well as individual historic buildings. Changes aim to respect protected landscape purposes and meet community needs. Village Management

Schemes are means of working with each community to deal with Conservation Areas and identify where physical improvements might take place.

Recreation and visitor management: The area's popularity for walking long preceded its establishment as a national park and pressure for the right for the public to roam freely over parts of the area led to bitter clashes between ramblers and landowners in the 1930s. Today, access agreements over large areas of moorland allow freedom to roam all year round, except for a brief period when parts of the moors are closed for grouse shoots.

The high population in cities surrounding the park has prompted the Board to pioneer new ways of dealing with large numbers of visitors. A total of 18.5 million visits are made annually.

Most of the park has a network of roads, tracks and paths set in landscapes of farmland, woodland and villages with recreation facilities of modest scale linked to features of interest and to riding, walking and cycling. There are 6000 km of public footpaths, and recreational zoning tries to ensure that the landscape can absorb appropriate use.

For vehicles, the 'Routes for people' concept has been introduced with cooperation from highway authorities. It segregates heavy traffic from villages and tourist traffic. But problems remain in handling visitor volumes in such popular places as Dovedale and Castleton and on the long-distance footpath – the 400 km Pennine Way – which starts at Edale.

The Board places much emphasis on education, information, and interpretation. It manages a large residential National Park Study Centre, offering activity holidays to the general public, training courses for countryside professionals, and services to schools and other educational groups visiting the park. The Board runs four Information Centres; publishes many guide books and leaflets about the park; offers wayside interpretation in many places; and organizes lectures and other events.

Staffing and budget: The Board has a multidisciplinary team of 160 full-time staff, many seasonal assistants, and about 300 part-time or voluntary rangers, all under the direction of the National Park Officer. Its annual budget (1989–90) is £4.75 million of which 51% came from Government, 19% from constituent local authorities, and 30% from trading income.

Management agency: Peak Park Joint Planning Board, Aldern House, Baslow Road, Bakewell, Derbyshire DE4 1AE, UK.

Normandie–Maine, France

Assessment: As with other such areas in France set up from 1975, the Normandie–Maine Regional Nature Park was established by Ministerial decree but the initiative for its establishment came from the region. It operates under a plan or charter developed between the region and the local communities. This charter covers administrative organization, plan of work, park facilities, legal measures and arrangements to finance facilities and management. The charter guides policies but the park has no statutory power. Paradoxically, however, this can be seen as a benefit as it compels the park team to keep in close touch with local people, local councils and other bodies. It must convince them that the park's proposals can benefit their communities; having achieved this the park's policies can then be applied through the organizations with statutory powers. The formula to achieve protected landscape goals is one of partnership. In theory the formula is excellent, but the success rate is variable. Much depends on the type of project and the degree of public interest. Moreover, changing policies already in existence is not easy and involves a great deal of work and time on the part of the park which has only a very small professional team in comparison with the magnitude of the tasks involved. However, significant benefits to both the landscape and the residents are achieved.

IUCN category: V; protected landscape.

Location: In the Orne department of the Basse–Normandie region in north-west France, near Alençon and some 100 km west of Paris.

Physical characteristics: The landscape is one of narrow gorges and broad valleys mostly lying on the Armorican massif which consists of granite in association with Jurassic deposits. Altitude rises to 417 m.

Nature conservation values: Vegetation includes numerous wealden-like wooded areas of oak, beech and pine in association with fir. The largest woodlands are the Forêts des Ecouves and Perseigne with deep glades of deciduous trees along with pine and spruce. Fauna includes numerous woodland species such as red deer, squirrel and wild boar.

Cultural values: The area maintains evidence of the 'bocage' landscape (fields enclosed by a network of hedges and trees) which was, until 1960, a functional part of the farm production system with wood used for heating, cooking, building and tool making. The area was important during later campaigns of the Second World War and various monuments commemorate the exploits of the allied forces.

Human population: Some 90 000 people in the park in 143 communes within the Orne, Mayenne, Sarthe and Manche departments.

Economic activities: Collapse of fuelwood-related industries (such as forges) after 1914 saw a drastic decline in population. This left an economy based on small-scale agriculture, polyculture, cider-making, cattle breeding (Normandy, frisonne pie noire) and rye grass hay meadows.

Establishment: The area was established by Decree as a 'regional nature park' in 1975 to maintain and preserve the traditional landscape and lifestyle of the region and stimulate local enterprises and rural development along with facilities for tourists. The park was established on the initiative of the local population, local authorities, trade and industry and regional associations.

Area: 234 000 hectares.

Land tenure: Private and State ownership.

Local administration: The park has a Committee of Management representative of the national forestry office, universities, municipalities and professional organizations, with a park director. The committee has no regulatory powers as most planning-related decisions are made and implemented by local government. Dialogue is therefore fostered between the park and local councils and other organizations with specialist committees set up to work in partnership with local interests to achieve protected landscape rural development goals.

Land use issues: Major issues relate to pressures for change and include removal of hedges and threatened loss of the characteristic 'bocage' landscape and the widening and straightening of water courses to improve stream flow and assist drainage of land for agriculture. The park has a special development plan aimed at maintaining and preserving the traditional landscape which is codified in a charter accompanied by a budget for investment and operating cost.

Conservation management: The park administration works at two levels to foster environmental and landscape conservation. First, through public awareness programmes and briefing those involved in rural management projects. Second, by disseminating conservation ideas locally through the technical and professional staffs of other organizations involved in countryside management. It fosters partnerships/committees bringing interests together. For example, after a park-sponsored survey of the impact of drainage schemes on stream flora and fauna had been completed, the park established a committee including the department responsible for river work as well as environmental, recreation and scientific interests with the aim of influencing future policy and practice.

Loss of the region's characteristic hedges prompted a project:

1. to make people more aware of the usefulness of hedges through teaching in agricultural schools and training teachers;
2. to take hedges into account in land management, working with the local land reorganization commission and providing conservation training for its members;
3. to support hedge planting and maintenance by helping to establish a woodland association, providing and demonstrating relevant techniques and equipment;

4. to illustrate the economic value of hedges by producing a feasibility study for a collective hedge management programme; compiling an inventory of public buildings (schools, public halls) which could use wood for heating; demonstrating economic advantages of retaining hedges in terms of income from a 'crop'; harvesting for fuel, woodchips for hardboard, timber; and establishing a cooperative to enable farmers to create an economic 'hedge wood system' from production to consumption.

Many other activities of the park authorities are orientated towards assisting local industries, such as restoration of wood industries. The park has helped establish cider and pear works and has established a biannual forum on wood and forest management. Wildlife conservation is incidental and there is no ban on hunting.

Human settlements: Traditional architecture is encouraged. The park authorities have assisted in restoration of forest crafts and up to half the population of some villages are now employed in timber trades.

Recreation and visitor management: The Maison du Parc at Carrouges supplies information on lifestyles, history and environment of the region as well as audiovisual displays and exhibitions. There are also a number of museums, woodland craft presentations, information centres and exhibition halls. The park assists active local groups particularly when their priorities coincide with those of the park. For instance, the park together with the Regional Group of Associations for Conservation runs courses for leaders of hiking and rambling groups to help them discover and appreciate the value of nature in the countryside while enjoying recreational activities.

Staffing and budget: The cost of facilities is borne by local communities with state subsidy while operating costs are shared among the constituent authorities.

Management agency: Normandie–Maine Parc Naturel Régional, Maison du Parc, BP 05, 61320 Carrouges, France.

Jurassic, Poland

Assessment: The Jurassic Landscape Parks (Zespo) (Jurajskich Parkow Krajobrazowych) (Plate 9) make up a protected landscape complex with a core national park, all surrounded by buffer zones. It is a good example of a multiple-zoned landscape area incorporating both protective and recreational features, as applied in a macro land-use planning model. Its administrative division into three (later four) voivodeships is an example of how regional authorities can cooperate on a complex land-use conservation system. However, it demonstrates that no protected landscape is an island as its major conservation problem, atmospheric pollution, originates from well outside its boundaries (Plate 10).

IUCN category: V; protected landscape.

Location: In south-central Poland in a triangle of land between the towns of Czestochowa in the north-west, Katowice in the south-west and Krakow in the south-east; close to the Upper Silesian industrial region, the industrial heartland of Poland.

Physical characteristics: Limestone rock and pinnacle outcrops, karst elements with extensive cave systems, ruins and remains of castles and Middle Ages fortifications, and scattered seminatural vegetation are the key characteristics. They follow the crest of the Jurajski–Czestochowa Uplands and include numerous valleys cutting deeply into the limestone plateaux. Altitude is 500 m. Average annual temperature is 7.5°C, with –3.5°C in January and 18°C in July. Rainfall increases from the north (550 mm) to the south (750 mm).

Nature conservation values: The area is now largely agricultural, with forest restricted to the steeper slopes and plateaux valleys. The limestone cliffs and pinnacles were at one point the only open areas supporting grassland communities. Forest areas cleared for timber have subsequently formed secondary grassland communities. The forests are largely pine in sandy areas, with beech elsewhere. Areas of oak, hornbeam and mixed pine-broadleaf are rarer. Meadows are very restricted in distribution but support a rich flora typical of lowlands, but with elements of mountains, as well as specialist species restricted by climate, humidity and soil structure. A number of endemics occur. The parks, with their cave system, are best known for their bats, with 17 out of Poland's 20 species present. Bird species include stone curlew, eagle owl, lesser grey and red-backed shrike, rock thrush and grey-headed woodpecker. Amphibians and reptiles and several butterfly species are present and molluscs are represented by typical mountain species.

Cultural values: There is evidence of human occupation back to the Ice Ages in remains in caves and burial plots, as well as ancient cemeteries. Ruins of castles and forts on the uplands of the 'Eagles Nest' testify to the importance of the area in historical times. The geographical lie of the uplands became a natural defence line in the Middle Ages for the Silesian region. Many events in Poland's history took place in this region, being situated near the ancient capital, Krakow. The more valuable and better preserved features are being restored.

Human population: No figures are yet available.

Economic activities: The predominant activity is agriculture, followed by forestry. Large areas are damaged by open-cast mining of minerals and rocks, and widespread extraction of sand has worsened the water balance by creating artificial sinks. Numerous, usually small, industrial installations exist, including cement works, paper factories and mines.

Establishment: The Jurassic Landscape Parks complex consists of seven landscape parks in a 90 km by 45 km block of upland and one separate unit some 50 km further to the north-west. The complex is zoned: the core is Ojcow National Park surrounded by a buffer zone which adjoins the Dolina Krzkowskic Landscape Park to the south and west. Further out from the core area are another six landscape parks. The national park was created by order of the Council of Ministers in 1956; the landscape parks and the buffer zone were established under orders of three different voivodeship people's councils. The outlying unit, including Zaleczanski Landscape Park, covering 7.1 ha and a buffer zone of 7.2 ha, was created by Sieradz voivodeship on

5 January 1978 and is proposed for inclusion in the complex. In Katowice voivodeship on 20 June 1980 an area of 73.8 ha was created, comprising a landscape park (29 ha) and buffer zone (44.8 ha); in Krakow voivodeship on 2 December 1981 94.8 ha was created, consisting of a landscape park (37.2 ha) and buffer zone (57.6 ha); and in Czestochowa voivodeship on 17 June 1982 77.6 ha was created comprising a landscape park (37.6) and buffer zone (39.9 ha).

Area: 246.3 hectares of which the parks cover 103.9 hectares and the buffer zones 142.4 hectares. Ojcow National Park covers a further 1509 hectares.

Land tenure: Mixture of state-owned and private. Most forest areas are state-owned whereas most agricultural holdings are private.

Local administration: Ultimate responsibility lies with the individual voivodeship authorities within which the parks lie. An overall coordinating committee consists of representatives from all the landscape parks and the three voivodeships. None of the individual parks has its own administration.

Land use issues: The area has had human occupation for thousands of years. Earlier there was pressure on the forests from farmers and later, as industry flourished, the remaining forests supplied timber for the nearby steelworks. This deforestation process expanded rapidly in the 19th century with a noticeable change in species composition. The small areas of peat bogs have been drained. Current pressures are from industry (industrial atmospheric pollution arriving on the westerly winds make the area some ten times as polluted as other areas), domestic construction and tourism.

Conservation management: The first nature conservation orders were issued to protect the seminatural forests at Zloty Potok and in the valley of the river Pradnik, actions which resulted in the creation of Ojcow national Park. Within the present-day park borders there are twenty nature reserves (covering 680 ha), including eight landscape reserves, six forest reserves and three floral reserves. In addition, there are many natural monuments as well as a number of rural parks on the site of country houses, such as at Pilicy and Zloty Potok. Almost the entire area is threatened by industrial pollution, some parts gravely. The sources are at a distance from the park; and no obvious solutions have been presented.

Human settlements: There are no large towns within the parks or its buffer zone, but numerous sizeable settlements and villages exist.

Recreation and visitor management: Being within striking distance of Poland's industrial heartland and the agglomerations of Katowice, Czestochowa and Krakow, tourist pressure on the parks is enormous. The southern parts are particularly heavily used and visited, in the form of both motorized and pedestrian tourism. Single day visits predominate. The long-distance footpaths, originally created in the 1970s, running the entire length of the uplands (one 168 km and the other 161 km long), are badly serviced with overnight accommodation and food outlets. There are, however, a number of individual sites well organized to accept tourists, such as at Pieskowa Skola and Ogrodzieniec. Each landscape park has its own network of camps, local routes and some facilities. Specialist activities include rock climbing, and a route has been created which links together all cultural features.

Staffing: Not separately staffed.

Management agencies: The Jurajski Landscape Park as it is known is administered by one coordinating authority and five vovoideship authorities. The coordinating authority is:
Zarzad Zespolu Jurajskich Parkow Narodowych, 31–227 Krakow, ul. Bagockiego 1a, Poland.
The vovoideship authorities are:
Bielsko Biala, Zywiecki Park Krajobrazowy, Bielsko Biala, ul. Sobieskiego 105, Poland.
Czestochowa, ul. Zaredzka 19, woj. Czestochowskie, Poland.
Katowice, 41–Dabrowa Turnia, ul. 27 Stycznia 69, Katowice, Poland.
Krakow, 31–227, ul. Bagockiego 1a, Sieradz, ul. Polna 27, Poland.

Appendix 1
International cooperation for promoting the concept of protected landscapes and seascapes (Resolution 17.43 of the 17th General Assembly of IUCN)

Resolution 17.43 of the 17th Session of the General Assembly of the International Union for Conservation of Nature and Natural Resources (IUCN) was adopted during the General Assembly held from 1–10 February 1988 in San José, Costa Rica.

With it, the highest decision-making organ of IUCN recognized 'the great value of the management category of Protected Landscape' and urged national and international action to promote the concept.

NOTING that the majority of governments now recognize the necessity to link the conservation of natural resources with economic development, following the basic principles of the World Conservation Strategy and the report of the World Commission on Environment and Development;

RECOGNIZING that while Strict Nature Reserves and National Parks (IUCN Categories I and II) contribute to conservation and economic development through non-consumptive uses of natural resources, and conserve areas of natural habitat with minimum human influence, no single approach to conservation is sufficient;

REALIZING, in this connection, that areas where people are a permanent part of the landscape can demonstrate durable systems of use that provide economic livelihoods, are socially and spiritually satisfying, are in harmony with nature, and preserve the cultural identity of communities;

REALIZING FURTHER that:

1. landscapes that have been materially altered by human activities often include species and ecosystems that are dependent on such activities;
2. such landscapes can serve as buffer zones of more strictly protected areas;
3. they can provide for recreation and tourism, which can make an important contribution to the physical and mental health of visitors as well as help develop public support for environmental protection;
4. such areas can form the basis for sustainable development over relatively large regions and thereby be of particular importance in many developing countries;

ACKNOWLEDGING:

1. the great value of the management category of Protected Landscape (IUCN Category V) for controlling inappropriate land uses and development pressures in outstanding human modified landscapes;
2. the value of the Biosphere Reserve concept in linking human concerns with those of protected areas;
3. the specific mention in the World Heritage Convention's Operational Guidelines of the value of areas with significant combinations of cultural and natural features;
4. the points made in the Lake District Declaration, which was unanimously adopted by the Symposium on Protected Landscapes held in the United Kingdom in October 1987;

the General Assembly of IUCN, at its 17th Session in San José, Costa Rica, 1–10 February 1988:

1. RECOMMENDS that, within the resources available, the Director General of IUCN should:
   (a) encourage IUCN members having experience and expertise in the establishment and management of protected landscapes and seascapes to make such expertise widely available to other IUCN members (perhaps by using IUCN's Commissions on National Parks and Protected Areas and on Sustainable Development as conduits);
   (b) assign the Commission on National Parks and Protected Areas to:
      (i) critically evaluate, develop and promote further the criteria for Category V: Protected Landscapes and Seascapes;
      (ii) develop management regimes for such areas;
      (iii) work with the United Nations Educational, Scientific and Cultural Organization (UNESCO) and the International Council on Monuments and Sites (ICOMOS) to develop criteria for the consideration of sites with mixed cultural and natural values for the World Heritage List;
   (c) request IUCN's Commission on Environmental Policy, Law and Administration, within the resources available, to conduct a survey of legal regimes applicable to the establishment, management, and

administration of protected landscapes and seascapes, and publish guidelines for establishing and implementing legal, administrative, and fiscal measures appropriate to the circumstances in different countries;

(d) actively promote the work of IUCN's Conservation Monitoring Centre to maintain data files on all categories of protected areas, with a particular effort directed to improving databases of those categories neglected to date, and to developing simple software which will enable governments and local management authorities to maintain their own compatible databases on personal computers;

(e) encourage IUCN's Conservation for Development Centre to work with governments and development assistance agencies to find ways and means to provide effective support to all categories of protected areas in developing countries as a concrete measure to harmonize conservation and development, and to ensure that the concept of different categories of protected areas is fully incorporated in any national conservation strategy that IUCN may be supporting;

(f) explore, via IUCN's programme in the regions, the application of ecodevelopment techniques in the sustainable use of protected landscapes.

2. FURTHER RECOMMENDS that governments and their agencies should:

(a) examine their systems of protected areas and other conservation measures, and develop, where necessary, designations and legal regimes for categories of protected areas that include people living permanently within the boundaries of the area;

(b) encourage the World Heritage Committee to adopt the principle that selected protected landscapes possessing significant harmonious associations of cultural and natural features can be considered as being of outstanding universal value and worthy of inscription on the World Heritage List;

(c) support other international efforts, such as the Action Plan for Biosphere Reserves and the Convention on Wetlands of International Importance especially as Waterfowl Habitat (Ramsar Convention), which promote effective management of protected landscapes in ways which respond also to the needs and aspirations of resident populations;

(d) promote, in particular through the Council of Europe and the European Federation of Nature and National Parks, the establishment of an International Seminar on Protected Areas in Europe, as an effective means of two-way transfer of knowledge about how to manage areas of outstanding conservation value that contain resident human populations;

(e) develop further ways and means for ensuring that people who live in and around protected landscapes are encouraged, with incentives where appropriate, to maintain a harmonious balance with the environment.

Appendix 2
Categories of protected areas

The rationale for a broader range of categories discussed in Chapter 1 of this book was explained in the IUCN report (1978). This report states that 'Management Categories can be designed and implemented so that each addresses a compatible set of benefits, without the pursuit of any one benefit ruling out the possibility of receiving other benefits'. It adds that 'even among generally compatible activities, conflicts may arise. But these can be treated through a zoning system or a periodic restricted activity system.'

The IUCN report said that areas which are managed to meet specified conservation objectives can be considered to be 'protected areas' and can be classified according to the objectives for which they are managed. However, the 'specific means required to meet the objectives of conservation will depend upon each particular situation and will vary with cultural, institutional, political and economic considerations'.

Anticipating what was to be the central goal of the World Conservation Strategy, and the WCED report, IUCN said that 'conservation categories provide the basis for clearly incorporating conservation into development (eco-development) ... Viewed in this way, conservation categories become means for sustained development.' The IUCN report concluded that, taken together, the categories can be administered as a unified system of conservation areas. Particularly significant was the recognition of the value within such a system of the 'lived-in' landscape which became identified as Category V Protected Landscape and Seascape.

The end product of the 1978 report was a decision by CNPPA to recognize ten categories divided into three groups. IUCN sees the various categories as meeting different conservation goals in different situations and Dr Kenton R. Miller, when Director General of IUCN in 1987, urged that people 'drop once and for all the perception of a hierarchy among categories of management'.

Using the definitions from the 1978 report, McNeely and Miller (1983) of IUCN listed two groups of protected areas covering eight categories, as well as identifying two internationally recognized designations – Biosphere Reserves and World Heritage Sites (Natural). Subsequently, a third such designation was added of Wetlands of International Importance (Ramsar sites).

Group A

Those categories for which CNPPA and WCMC take responsibility to monitor the status of each conservation area and for which CNPPA takes a responsibility to provide technical advice as requested.

Category I: Scientific Reserve/Strict Nature Reserve

To protect nature and maintain natural processes in an undisturbed state in order to have ecologically representative examples of the natural environment available for scientific study, environmental monitoring, education, and for the maintenance of genetic resources in a dynamic and evolutionary state.

Category II: National Park

To protect natural and scenic areas of national or international significance for scientific, educational, and recreational uses.

Category III: Natural Monument/Natural Landmark

To protect and preserve nationally significant natural features because of their special interest or unique characteristics.

Category IV: Nature Conservation Reserve/Managed Nature Reserve/ Wildlife Sanctuary

To assure the natural conditions necessary to protect nationally significant species, groups of species, biotic communities, or physical features of the environment, where these require specific human manipulation for their perpetuation.

Category V: Protected Landscape or Seascape

To maintain nationally significant natural landscapes which are characteristic of the harmonious interaction of People and Land, while providing opportunities for public enjoyment through recreation and tourism within the normal lifestyle and economic activity of these areas.

Group B

Those categories important to IUCN as a whole and generally found in most nations, but not be considered exclusively within the scope of CNPPA.

Category VI: Resource Reserve – Interim Conservation Unit

To protect the natural resources of the area for future use, and prevent or contain development activities that could affect the resource pending the establishment of objectives which are based on appropriate knowledge and planning.

Category VII: Natural Biotic Area/Anthropological Reserve

To allow the way of life of (human) societies living in harmony with the environment to continue undisturbed by modern technology.

Category VIII: Multiple-use Management Areas/Managed Resource Areas

To provide for the sustained production of water, timber, wildlife, pasture, and outdoor recreation, with the conservation of nature primarily oriented to the support of economic activities (although specific zones may also be designed within these areas to achieve specific conservation objectives).

Appendix 3
Area of outstanding natural beauty

Proposal for a protected landscape

A leaflet to inform interested parties about the implications of a proposal for a protected landscape in the Ring of Gullion (Plate 11) area of Northern Ireland is an excellent example of an important part of the process of public consultation. The leaflet was prepared by the Department of the Environment for Northern Ireland.

In its original form, the leaflet was attractively illustrated and had an excellent map showing proposed boundaries.

About this leaflet

This leaflet is your invitation to comment on the proposal to designate a Ring of Gullion Area of Outstanding Natural Beauty (Plate 11).

The Department of the Environment considers that the area around Slieve Gullion in South Armagh merits special attention because of its outstanding landscape character. The Department is seeking your views on the proposal.

What is an Area of Outstanding Natural Beauty?

An Area of Outstanding Natural Beauty (AONB) is an area of countryside with exceptional qualities of landscape, heritage and wildlife. An AONB is designated by the Department as part of the process of protecting and conserving these qualities and promoting their enjoyment. It is intended to be a positive contribution to rural development.

Why should the Ring of Gullion be an AONB?

The Ring of Gullion is a unique geological feature which has its own distinctive landscape character unparalleled elsewhere in the British Isles. Formal designation as an Area of Outstanding Natural Beauty is recognition that the countryside of Slieve Gullion and its surrounding ring of hills are national assets.

The special character of the area includes:

1. the heather clad mountain of Slieve Gullion and the steep and craggy hills forming the surrounding ring;
2. the remarkable geological history of an extinct volcano;
3. the rich wildlife of small areas of heath, mire, bog and woodland within the farmed countryside;
4. the fascinating historical heritage with a concentration of prehistoric stone monuments and Early Christian sites including Ballykeel Dolmen, The Dorsy and Killevy Churches;
5. the strong visual impact of neatly patterned fields and ladder farms contrasting with patches of woodland and bog or heather hillsides;
6. the traditional character and distribution of older farmsteads and their associated gate pillars and wrought iron gates;
7. the attractive Camlough Lake, important for wildlife and angling;
8. the opportunities for informal countryside recreation at picnic sites and viewpoints and in forest areas; and
9. the folklore and ancient myths associated with the Slieve Gullion countryside.

What is the purpose of designation?

The purpose of designation is to provide a framework within which the Department may agree policies and proposals for:

1. conserving or enhancing the natural beauty or amenities of the area;
2. conserving wildlife, historic features or natural phenomena within it;
3. promoting its enjoyment by the public; and
4. providing and maintaining public access to it.

These policies and proposals will be drawn up in consultation with the people of the area, Newry and Mourne District Council and the Department's Advisory Committee, the Council for Nature Conservation and the Countryside.

Within the AONB other public bodies in the exercise of their statutory responsibilities will be obliged to give special consideration to the conservation of the countryside.

BUT the designation will NOT affect the ownership or occupation of the land and will NOT interfere with the duty and role of the District Council in the administration of the area.

What does the department do in an AONB?

The functions of the Department include:

1. grant-aiding district councils and other organizations in the provision of facilities for countryside recreation such as open spaces, heritage centres, nature trails, footpaths and public rights of way;
2. advising, assisting and where appropriate making agreements with farmers who wish to manage their land with regard to conservation principles for wildlife and public enjoyment;
3. protecting wildlife in the countryside and in nature reserves or other sites of special value and guarding against the loss of endangered species or habitats;
4. providing funds for voluntary organizations working in the countryside;
5. monitoring environmental pollution and taking action against offenders;
6. recording, protecting and advising on historic monuments and buildings so that they may be enjoyed by future generations;
7. providing information such as guides, booklets and illustrated material on the countryside for the benefit of school groups and the public;
8. ensuring that development is planned in an orderly and efficient way and does not detract from the quality of the countryside;
9. requiring high standards of location, siting and design for new dwellings and encouraging traditional building styles;
10. conducting research on matters affecting the countryside, its economy and its continuing enjoyment by locals and visitors, and suggesting ways in which the economic viability of the area be maintained and enhanced; and
11. advising other government agencies responsible for agriculture, forestry, communications and tourism.

Have you any comments?

If so, they should be sent in writing to:

Department of the Environment (NI)
Countryside and Wildlife Branch
Calvert House
23 Castle Place
Belfast BT1 1FY

All letters with comment should be marked 'Ring of Gullion' and be received before 2 July 1990.

Detailed maps of the proposed boundary are on display in the District Council Office in Newry and in local libraries and community centres.

What is the next stage?

When the Department has received your response it will consider whether to proceed with designation of the boundary as proposed. If so, a formal Designation Order will be published as prescribed in Part IV of the Nature Conservation and Amenity Lands (NI) Order 1985.

In conjunction with the people of the area, the District Council, and other statutory and voluntary bodies the Department will agree more detailed policies and proposals for countryside management. In this way the Department hopes to conserve the landscape and to promote enjoyment of the countryside in the Ring of Gullion.

Appendix 4
A management planning process

Introduction

It is a basic principle of protected area management, that every protected area should have a management plan. The management plan helps identify key issues, states policies to achieve objectives, sets priorities and details strategies to implement them. With a consultative planning process, it helps form or maintain a partnership between public and private interests.

Central to such a plan is a statement of goals and measurable objectives to guide the management of the area. These goals and objectives form the framework for determining what actions to take, when they will be taken, and the budget and personnel needed to implement them. A management plan is a valuable tool for identifying management needs, setting priorities and organizing the approach to the future.

A management plan provides this guidance for a specified period of time, typically five years. Annually operations plans may be developed during the implementation phase using the management plan as a guide. The management plan is always subject to modification as new information is obtained, particularly through feedback on the effectiveness of actions taken under the annual operations plan.

By identifying the management steps necessary for the protected area, and the resources needed to take them, the management plan helps the manager to allocate, and make best use of, staff, funding, equipment and materials. Where resources are inadequate to achieve the management objectives, the plan can be used to document these deficiencies and list the area's needs. In this way the plan becomes a valuable tool to seek needed support.

A management plan also serves as a communication tool to gain the understanding and support of both the general public and relevant officials. Such understanding is important for gaining the cooperation of local people and the political support needed for adequate funding.

Finally, the management planning process can be important training for management personnel. Involvement in the planning process exposes staff to the full range of management needs and to the public and other interests and this leads to a fuller understanding of their role. The management plan aids this process by providing continuity over time and facilitates consistency during staff transfers.

Although planning methodology can be complex, the basic steps in an idealized procedure are outlined below. The 16 steps listed cover the full range of possible factors to be considered. Needs, limitations and priorities will vary widely with each situation and the management planner must tailor the process to meet the particular circumstances.

Although the planning process is summarized as a list of steps, it will often be necessary to consider some steps simultaneously, and to review earlier decisions as new information becomes available.

As a final caveat, it should be noted that the following steps display a thought process, a means of organizing a future based on an assessment of the present. It is not, therefore, the nomenclature or the exact order which is important but the process by which one evaluates and addresses the management problems of a protected area.

Step 1: Set up the planning structure

Although a plan can be prepared by one individual, it is most often a team exercise involving a core group of three to six individuals. It is useful if the team members have a mix of capabilities in the methodology of planning, ecology, sociology, economics, and various other resource sciences.

The preparation of a management plan also requires the participation of those who now manage the area as well as those who use the area or who will be affected by the plan.

A useful approach with a protected landscape would be for the planning team to work to the local protected landscape body through a small committee with representatives of landowners, local communities, local government and user groups to serve as a consultative body for the planning team and maintain citizen involvement. Members should be knowledgable about the area, concerned about its conservation and able to work easily with other people.

Step 2: Gather basic background information

The next step is a review of all resource material available on the protected area. This includes the enabling legislation, data on biophysical features, cultural resources, and socioeconomic data. This information can be collected from various sources including a literature review, office files, and interviews with knowledgeable people. A base map and reference file system are prepared at this stage.

This material is essential as a basis for the plan but it is not desirable or necessary for the plan to attempt to be a compendium of all available information. What is important is the interpretation of key information relevant to management issues.

Step 3: Field inventory

Planning usually requires fieldwork to gather new information, to check and update existing data, and to view the area with new perspectives. The purpose is to develop the information base needed to make informed management decisions. Generally, a review is made of environmental resources and use by residents and visitors. Attention is given to archaeological sites and contemporary cultures, regional economics, transportation networks, and attitudes and needs of local people. Particular attention is devoted to critical areas. Field staff play a key role in accumulating the necessary data.

Again, it is important to remember that information is collected to identify the most important management needs, not as an end in itself. Collect only that information which is most pertinent to management.

Step 4: Assess limitations and resources

Limitations of an environmental, economic, political, administrative or legal nature should be recognized and analysed at this point. Senior management plays a critical role in defining these limitations, and identifying key problems of the site. A review of the implications of national policies or plans and relevant regional or district development plans should be done at this stage. Aim to ensure that options to be developed in later stages will be realistic in the context of these policies and plans. Resources should be recognized and analysed to make effective use of them.

Step 5: Review regional interrelationships and develop a consultation process

A protected landscape must be integrated as an essential element in the regional land-use pattern. The planning team must attempt to review the potential effects of development outside the protected landscape boundaries as well as the effects of the protected landscape on the region. A consultation process to involve local government, organizations and residents needs to be developed and implemented as planning proceeds to ensure that these key interests are involved in the process.

Step 6: State the objectives of the area

With the above steps completed, it will be possible to spell out in detail the values and objectives of the area in relation to its particular set of resources, to the region, and to the country as a whole.

Step 7: Divide the area into management zones

Most protected areas will be zoned for different objectives and uses and/or may be broken for planning purposes and consultation into geographic regions.

Step 8: Review boundaries of the area

Few protected landscapes will have ideal boundaries. With the resource inventory, management objectives, regional integration review, and zoning stages of the plan in place, the team should now consider if there is a case for boundary modifications.

Step 9: Design the management programmes

Once the zoning concept has provided the basis for what is to be done where, the task now is to answer the questions How? and Who?. This action-oriented component is the heart of the plan and addresses the four major programmes of management.

1. Resource management and protection. This management programme focuses on issues relating to the protection of the biological and physical resources of the area.
2. Human use. This programme deals with all aspects of use by people including traditional use, recreation, tourism, interpretation, and extension, and the facilities and developments necessary for these purposes.
3. Research and monitoring. The management of protected area resources requires an understanding of specific ecological processes and protected landscapes require understanding and appreciation of the socioeconomic needs of the residents. One important aspect of management involves the design and development of research programmes to meet these needs. Similarly, a monitoring programme is needed to detect problems as they arise and gauge progress in meeting the management objectives of the area.
4. Administration. The operational, human and financial resources needed to manage the protected landscape are outlined. Headquarters facilities, vehicles, equipment, and maintenance requirements are some issues of concern here.

Step 10: Prepare integrated management options

This step in the management planning process summarizes all the programmes and physical facilities needed to achieve the various management goals. The team may wish to present various options.

Step 11: Outline financial implications

No plan can be evaluated without costing the planning proposals, at least as general estimates. In some cases, the economic justifications will need intensive treatment in a cost/benefit analysis. In any case, the planning team must now present the cost estimates of the proposals.

Step 12: Prepare and distribute a draft plan

Before proceeding further, the team should seek comment and feedback on the proposals. A first draft of a plan should be compiled and distributed to the range of actors who are the key to the plan's success, both within and outside the agency.

Step 13: Analyse and evaluate the plan

After digesting input from all concerned parties, the team is now in a position to narrow its options. A review of all development proposals is made, considered by management and adopted by the relevant protected landscape body.

Step 14: Design schedules and priorities

As the plan is now finalized and ready to be put in motion, the team and management decide on the timing and priorities of each event.

Step 15: Prepare and publish the finalized Plan

With the appropriate approval, the plan is produced, published and distributed in a form suitable to reach a general audience, including landowners and local residents. Copies of the document should be given to political leaders, ministry officials, and related government departments, regional councils, international agencies, scientific personnel involved in research and monitoring, and to appropriate public interest groups. Launching the plan should involve public meetings to explain it more fully.

Step 16: Monitor and revise the plan

Plans need revision as new information becomes available and basic conditions change. Thus a five-year planning horizon is often used as a realistic time-span for a management plan. As a final step the plan should be reviewed at intervals which may take place more frequently than five years.
Source: Adapted from MacKinnon *et al.* (1986).

Appendix 5
Summary of management plan review (1988) of Peak 'National Park', UK

A summary follows of the key elements of a management plan, in this case a 1988 review of the plan for the Peak National Park (UK) which holds a Council of Europe Diploma for its work in protected landscapes management.

A summary of the draft plan was sent to every household and business in the area with a questionnaire inviting comment. Public meetings were held and written comments invited.

The plan outlines objectives and policies for management of the area over at least the next five years. It does this under key headings of conservation, recreation and rural development and then has a section covering implementation. This is followed by a section which translates the implications of the plan to specific areas of the protected landscape and key areas adjacent to it. This has the advantage of enabling all interested and, particularly residents, to appreciate the implications of the plan for the area of their particular interest.

The contents page of the plan contains the following headings.

Introduction

The basic legislation
Changes in circumstances and ideas
Format of the plan
Facts and figures

Conservation

A strategy for conservation
Landscape conservation and land management
Farming
Forestry
Nature conservation
Other land management
Cultural heritage

Towns, villages and historic buildings

Recreation

A strategy for recreation
Rights of way
Access to open country
Other active recreation
Visitor facilities
Traffic management
Recreational public transport
Visitor accommodation and attractions
Information, interpretation and education
Ranger service

Rural development

A strategy for rural development
Housing, jobs and services
Regional and national needs

Ways and means

Direct action
Joint action with others
Grant aid
Advice, information and research
Statutory influence
Consultation
Monitoring
Choosing priorities for action

Area summaries

This section takes elements from the plan and relates them to specific proposals for all of the areas within the whole protected landscape so that management and all interested can see what is proposed area by area.

It is helpful to elaborate on the sort of material included in some sample chapters, specifically the Introduction, Conservation and the implementation chapter entitled Ways and means.

A very brief summary of the main elements of these chapters follows.

Introduction

This contains the background to the plan, a brief outline of the management structure, linkages with other statutory planning, the importance of management, partnership and resources.

A summary of changes in circumstances and ideas is given as the plan is a review of an earlier one. There are references to environmental awareness, changes in recreation, changes in the economy, and interest in cooperative action among various public and private sector interests.

The plan goes on to introduce a zoning mechanism within the protected landscape (as distinct from zoning for overall rural land-use planning) to take account of different situations requiring different approaches to management. In this case three conservation policy zones are identified.

1. The Natural Zone where there has been only minor modification by human activities. In Peak District, land zoned 'natural' covers gritstone moors, principal limestone dales and limestone heath and hill. Natural qualities must predominate in future management.
2. The Rural Zone where human influence is much more obvious consisting of enclosed farmland and managed woodland.
3. The Settlements where conservation policy is to maintain each settlement's distinctive character with changes respecting protected landscape purposes and meeting the needs of the resident community.

This section of the plan also identifies areas needing special attention including those having their own statutory protection as nature reserves, sites of special scientific interest, historic conservation areas and ancient monuments.

Landscape types are identified as areas particularly important to conserve – in this case moorland, limestone heaths and hills and woodlands with appropriate policies for each.

This section of the plan concludes with a statement of the objective for conservation, in this case:

> To maintain and enhance the particular qualities of the (area) by careful recognition of the particular elements (and combination of elements) that give each part of the area its distinctive (protected landscape) qualities.

Then follows a statement of policies, for example, 'In the Natural Zone there will be a general presumption against obvious new man-made features and wild semi-natural conditions will be enhanced'.

A similar pattern follows throughout the plan with presentation of key facts, identification of issues, and statements of the relevant objective and policies. These are drawn together in a chapter dealing with implementation.

Ways and means

This chapter discusses implementation of the plan under the following headings. The summaries recorded here represent very brief summaries of what is a significant part of the plan translating the objectives and policies into proposals for action.

Introduction

The emphasis is on the fact that the plan is 'to be implemented by all who work, live in or have responsibilities in' the area drawing on whatever resources can be obtained – from Government grant, local authorities, grants from other sources, volunteers, job creation programmes and direct income from the public.

Direct action

The protected landscape management body's direct role relates to management of land it owns and provision of services and this section of the plan outlines the land it manages and the services it provides such as information centres and services, a study centre, caravan and camping sites, youth hostels, cycle hire services, publications etc.

A statement of relevant policies follows.

Joint action with others

This outlines a wide range of joint programmes involving the landscape management authority with others. Examples are Area Management Schemes (with water authorities, the National Trust, the Forestry Commission, county, district and parish councils and other bodies); public path programmes (with highway authorities, district and parish councils, the National Trust, other landowners, and voluntary bodies helping maintain footpaths); and Enhancement of Conservation Areas (with public utility boards, highway authorities, parish councils, etc.).

Reference is made to other cooperative activities such as Rural Development and Tourist Action programmes, and Integrated Rural Development Project, Community Action in the Rural Environment (CARE) and Moorland Restoration and Management.

Again policies on future work in these fields are listed.

Grant aid

Practices and policies are outlined for use of grant aid funds for such things as planting or management of woodlands, conservation works in local nature reserves, landscape enhancement of unsightly areas, pond restoration, repair of historic features and provision of public facilities.

Annual grants or payments to stimulate continuing action by others to serve protected landscape purposes are outlined for such things as support to voluntary organizations carrying out a range of conservation work, access agreements, and management agreements under a Farm Conservation Scheme to secure protection or active maintenance by farmers of heritage features such a drystone walls, flower-rich meadows or woodlands.

Advice, information and research

This section covers policy on advice to farmers on all aspects of farm conservation, advice to owners of historic buildings and to property owners to influence developments they propose. Reference is made to provision of architectural or landscape design services (on a recharge basis) to other authorities and management (on an agency basis) of woodlands.

Policies on commissioning or co-sponsoring research are covered along with updating and improving the information base on the area and making it freely available to assist protected landscape goals.

Statutory influence

This section deals with the way in which the management authority uses its statutory powers or formal rights of consultation to achieve its management objectives. These include giving or witholding permission for changes in land use, or construction or alteration of buildings, or mineral working or the imposition of conditions to ensure developments are acceptable.

Other fields covered are advising the Forestry Commission on plans of operation by owners of key forests and on applications for forestry or woodland grants; applications from farmers for Ministry of Agriculture grants and the design, siting and materials of agricultural and forestry buildings and roads.

Consultation

This covers consultation processes 'to oil the wheels of partnership'. It includes provision for a range of liaison mechanisms with local government, rural development programmes, and tourism, recreation, wildlife conservation, historic and archaeological, farmers' and tenants' organizations.

Monitoring

A commitment is made to maintaining a system for monitoring the implementation of the plan and reporting to all involved on steps needed to achieve the goals of the protected landscape.

Priorities for action

This is a list of the main priorities for future action drawn from the wide range of policies stated in the plan. It forms a basis for the preparation of continuing programmes for future management.

There is reference to the implications of changes in relevant legislation, administrative structures and national policies which affect or may affect the area.

A brief conclusion states the need to reflect in the plan these implications or changes while reaffirming the lasting values which prompted the designation of the area as a protected landscape.

A section then deals with the format of the plan and a statement on the process of consultation including mechanisms used to involve representatives of major industries, particularly agriculture and forestry, in the consultation process.

Reference is then made to basic information about the area, its values, its people and about visitors and recreation use.

NOTE: There is the choice of maintaining separate documentation in a database on the area or incorporating details in the plan itself. The fact that the database will have important roles independent of the plan favours the maintenance of a separate database with the plan identifying key elements from the database. This approach also has the advantage of reducing the bulk and cost of publication of the management plan.

Conservation

This first identifies key elements which make the protected landscape distinctive and of particular interest to people. In the Peak case, these are identified as:

1. distinctive unspoilt beauty resulting from a subtle blend of natural forces and human activities resulting in settlements in sheltered hollows and largely built from local stone; vegetation cover often self-sown, influenced by grazing and simple farming methods; field boundaries constructed of local sources of water power;
2. the geographical position of the area as uplands surrounded on three sides by more fertile lowlands resulting in the wildlife and culture of the area displaying a distinctive mixture of upland and lowland, northern and southern features;
3. the atmosphere of the area; its 'sense of tranquility, of escape from urbanization and contact with nature' together with its qualities of wilderness in the moorlands and sense of history.

Appendix 6
Protected areas management agencies

Address list

Prepared by the Protected Areas Data Unit, World Conservation Monitoring Centre and Commission on National Parks and Protected Areas, IUCN – The World Conservation Union

Throughout the world there is a wide range of agencies responsible for protected areas in many nations. These agencies may be in environment or agriculture ministries, or stand-alone organizations. They may be at national level or state/provincial level. Some agencies may also be quasi-government or non-government organizations.

Some of the agencies listed are responsible for protected landscapes as well as other categories of protected areas. All have a potential interest in the protected landscapes concept.

Information in the list is organized by country. However, some islands and territories, while being part of one country, are geographically separate and have their own identity. These are included in the list under their own geographical name rather than as part of the 'parent' country, and the addresses of the protected areas management agencies within these areas are given.

Where information has been available to the compilers, the title of the senior officer within each agency is included, followed by the postal address of the agency's headquarters. Telephone, telex and fax numbers are included where they are known, with the country codes for telephone and fax numbers shown in brackets.

The information has been compiled by staff of the World Conservation Monitoring Centre, using the extensive protected areas database managed by the Centre. The information has subsequently been reviewed by staff of the IUCN Protected Areas Unit in Gland, Switzerland and by officers of the IUCN Commission on National Parks and Protected Areas.

Any list of this kind is bound to include inaccuracies, inconsistencies, and outdated information. Corrections should be sent to:

Protected Areas Data Unit
World Conservation Monitoring Centre
219 Huntingdon Road
CAMBRIDGE CB3 0DL
United Kingdom

Tel: (44) 223 277314
Tlx: 817036 scmu g
FAX: (44) 223 277136

Afghanistan

Department of Forests & Range
Ministry of Agriculture & Land Reform
KABUL
Afghanistan

Tel: 408415

Albania

Ministry of Agriculture
TIRANA
Albania

Algeria

Director
Direction de la sauvegarde et de la promotion de la nature
Secrétariat d'Etat aux forêts et à la reforme agraire
B P 86
Kouba, ALGIERS
Algeria

American Samoa

Department of Parks & Recreation
USDI
P O Box 3809
PAGO PAGO
American Samoa
USA 96799

Andorra

Conselleria d'Agricultura
Edifici CASS 5e pis
Carrer Maragall S/N
ANDORRA LA VELLA — Tel: (33) 628 21234
Andorra

Angola

Ministerio da Agricultura
Direcção Nacional da Conservação da Natureza
Rua Eng° Artur Torres, 10—A/5° Esq
C P 74 — Tlx: 3322
LUANDA
Republica Popular de Angola

Anguilla

Ministry of Tourism, Agriculture and Fisheries
The Valley — Tel: (1) 809 497 2625
Anguilla

Antarctica

See Australia (Tasmania), France, Norway, South Africa, UK

Antigua & Barbuda

Fisheries Department
Ministry of Agriculture, Lands and Fisheries
Government Headquarters
ST JOHN'S — Tel: (1) 809 460 1007
Antigua & Barbuda — FAX: (1) 809 460 1516

Parks Commissioner
Antigua and Barbuda National Parks Authority
P O Box 1283
ST JOHN'S — Tel: (1) 809 460 1053
Antigua & Barbuda — FAX: (1) 809 460 1516

Argentina

Presidente
Administración de Parques Nacionales
Secretaria de Agricultura, Ganaderia y Pesca
Avda. Santa Fe 690 Tel: (54) 1 311 6633/8855/312 0257
1059 BUENOS AIRES Tlx: 21535 dgaag ar
Argentina FAX: (54) 1 111516

Australia

Director
Australian National Parks
and Wildlife Service
P O Box 636
CANBERRA CITY Tel: (61) 6 250 0222
Capital Territory 2601 Tlx: 62971
Australia FAX: (61) 6 250 0339/0228

Director
Australian Capital Territory
Parks & Conservation Service
P O Box 1119
TUGGERANONG Tel: (61) 6 293 5222
ACT 2690 FAX: (61) 6 293 5274
Australia

Director
National Parks and Wildlife Service
P O Box 1967
HURSTVILLE Tel: (61) 2 585 6300/6550
New South Wales 2200 Tlx: 26034
Australia FAX: (61) 2 585 6555

Director
National Parks & Wildlife Service
P O Box 1260
DARWIN
NT 0800
Australia

Tel: (61) 89 508211/815299
Tlx: 85130
Tel: (61) 89 813497

Director
National Parks and Wildlife Service
P O Box 190
NORTH QUAY
Queensland 4002
Australia

Tel: (61) 7 227 7111
FAX: (61) 7 221 7676

Director
National Parks & Wildlife Service,
P O Box 1782
ADELAIDE
South Australia 5001

Tel: (61) 8 216 7867
FAX: (61) 8 231 1392

Director
Department of Parks, Heritage & Wildlife
GPO Box 44A
HOBART
Tasmania 7001
Australia

Tel: (61) 2 302336
FAX: (61) 2 238765

Director
National Parks & Wildlife Division
P O Box 41
EAST MELBOURNE
Victoria 3002
Australia

Tel: (61) 3 412 4011
Tel: (61) 3 412 4119

Director
Parks, Recreation & Planning
Department of Conservation & Land Management
P O Box 104 — Tel: (61) 9 386 8811
COMO — Tlx: 94585 aa
Western Australia 6152 — FAX: (61) 9 386 1578

Chairman
Great Barrier Reef Marine Park Authority
P O Box 791
CANBERRA CITY — Tel: (61) 6 247 0211
ACT 2601 — Tlx: 62552 arric
Australia — FAX: (61) 6 247 5761

The Director
Antarctic Division
Channel Highway
KINGSTON, Tasmania — Tel: (61) 2 290209
Australia 7150 — Tlx: 57090

Austria

Bundesministerium für Gesundheit
und Umweltschutz
Stubenring 1
1010 VIENNA — Tel: (43) 1 222 66150
Austria

Bahamas

Bahamas National Trust — Tel: (1) 809 323 1317
P O Box N 4105 — (1) 809 323 2848
NASSAU N P — FAX: (1) 809 393 4978
Bahamas

Ministry of Agriculture, Fisheries &
Industry
P O Box N 3028
NASSAU N P — Tel: (1) 809 323 1777
Bahamas

Bahrain

Environmental Protection Committee
P O Box 26909 — Tel: (973) 293693/275792
ADLIYA — Tlx: 8511 Health BN
Bahrain — FAX: (973) 293694

Bangladesh

Conservator of Forests (General Administration & Wildlife)
Bana Bhawan
Gulshan Road
Monakhali — Tel: (880) 2 603537
DHAKA 12
Bangladesh

Barbados

Chairman
National Conservation Commission
P O Box 807E
Codrington House
ST MICHAEL — Tel: (1) 809 426 5373
Barbados — FAX: (1) 809 429 8483

Barbados National Trust
Ronald Tree House
2 10th Avenue
Belleville
ST MICHAEL — Tel: (1) 809 426 2421
Barbados

Belarus

State Committee for the Environment
Kollectornaya 10
220 084 MINSK
Belarus

Tel: (7) 172 206681
FAX: (7) 172 205583

Belgium

Administration de la recherche agronomique
Ministère de l'Agriculture
Av. du Blvd. 21 (7ème étage)
1210 BRUSSELS
Belgium

Tel: (32) 2 211 7323
Tlx: 22033
FAX: (32) 2 211 7216

Belize

Chief Forest Officer
Department of Forestry,
Ministry of Natural Resources
BELMOPAN
Belize

Tel: (501) 8 2333

Bénin

Directeur
Direction des forêts et des ressources naturelles
Ministère du développement rural
B P 393
COTONOU
République Populaire du Bénin

Tel: (229) 330662
FAX: (229) 33042

Bermuda

Conservation Division
Ministry of the Environment
30 Parliament Street
HAMILTON HM 12
Bermuda

Tel: (1) 809 2 55151

Bhutan

Ministry of Agriculture & Forests
Royal Government of Bhutan
P O Box 125
THIMPHU — Tel: (975) 2 2503
Bhutan — FAX: (975) 2 2395

Department of Forestry
P O Box 130
THIMPHU — Tel/FAX: (975) 2 2395
Bhutan

Bolivia

Jefe Nacional
Departamento de Vida Silvestre, Parques, Caza y Pesca
Centro de Desarrollo Forestal
Casilla de Correo 8124
Avda Comacho 1312
LA PAZ — Tel: (591) 2 374265
Bolivia — Tlx: 2697

Botswana

Director
Department of Wildlife and National Parks
P O Box 131 — Tel: (267) 51461/51790
GABORONE — Tlx: 2414
Botswana

Brazil

Secretaria Especial do Meio Ambiente (SEMA)
Secretaria de Ecosistemas
Coordenadoria de Areas de Proteção Ambiental
CP 70750
Av. W/3 Norte, Quadra 510, Lote 08 — Tel: (55) 61 274 9685
BRASILIA DF — Tlx: 611429
Brazil

Chefe do Departamento de Unidades de Conservação
Directoria de Ecosistemas
Instituto Brasiliero de Meio Ambiente e Recursos Naturais Renováveis
Sain Av. L4 Norte - ED
IBDF/MINTER
CEP 70800
BRASILIA DF
Brazil
Tel: (55) 61 223 7879/0901/321 2324
Tlx: 614304
FAX: (55) 61 224 5206/322 1004

Director
Depto. Parques Nacionales y Reservas Equivalentes
SBN 13° andar
70.000 BRASILIA DF
Brazil
Tel: (55) 61 225 8125
Tlx: 1171

Departamento de Recursos Naturais da Fundação Zoobotanica
Ed. Rogério Pitthon Farias
S. de Areas Isoladas Norte
70.000 BRASILIA DF
Brazil

Secretaria de Agricultura
Rua Raimundo Nonato 116
Forte São João
29.000 VITORIA
Espirítu Santo
Brazil

Instituto Estadual de Florestas
Rua Espíritu Santo 604 - 3° and.
30.000 BELO HORIZONTE
Minas Gerais
Brazil

Instituto de Terras e Cartografia
Rua Desembargador Motta 3384
80.000 CURITIBA
Paraná
Brazil

Departamento Geral de Recursos Naturais Renováveis
Secretaria de Estado de Agricultura e Abastecimento
Av. Marechal Câmara 314
20.020 RIO DE JANEIRO
Brazil

Fundação Estadual de Engenharia do Meio Ambiente
Rua Fonseca Teles 121, 14° and.
20.940 RIO DE JANEIRO
Brazil

Depto de Preservação e Controle de Recursos Naturais Renováveis
Secretaria da Agricultura
Av. Julio de Castilhos 585/5°
90.000 PORTO ALEGRE
Rio Grande do Sul

Fundação de Amparo à Tecnologia e Meio Ambiente
Praça Pereira Oliveira 14 – 2°
88000 FLORIANOPOLIS
Santa Catarina
Brazil

Instituto Florestal
Coordenadoria da Pesquisa de Recursos Naturais
Secretaria de Agricultura e Abastecimiento
Centro Estadual da Agricultura
Av. Miguel Estefano 3.900
04301 AGUA FUNDA
São Paulo
Brazil

British Virgin Islands

Conservation & Fisheries Department
Ministry of Natural Resources
Road Town
TORTOLA
British Virgin Islands
Tel: (1) 809 494 5681/2

Director
National Parks Trust
c/o Ministry of Natural Resources
Government Headquarters
Road Town
TORTOLA
British Virgin Islands
Tel: (1) 809 494 3904

Brunei Darussalam

Director of Forestry
Forestry Department
Ministry of Industry & Primary Resources
BANDAR SERI BEGAWAN 2067
Brunei Darussalam
Tel: (673) 2 240232/226493/222450
Tlx: 2228 contel bu
FAX: (673) 2 241012

Bulgaria

Ministry of the Environment
ul. Vladimar Poptomov 67
1000 SOFIA
Bulgaria
Tlx: 22145
FAX: (359) 521634

Ministry of Forests and Protection
of the Natural Environment
ul Antim I 17
1000 SOFIA
Bulgaria

Burkina Faso

Directeur des parcs nationaux, des réserves de faune et des chasses
Ministère de l'environnement et du tourisme
B P 7044
OUAGADOUGOU 03
Burkina Faso

Tel: (226) 3 32477
Tlx: 5283
FAX: (226) 3 311142/301351

Burundi

Directeur général
Institut national pour l'environnement et la conservation de la nature
B P 56
GITEGA
Burundi

Tel: (257) 40 2071
Tlx: 3000 Gitega
FAX: (257) 40 2075

Cambodia

Service de Forêt et Chasse
Ministère de l'Agriculture
PHNOM PENH
Cambodia

Cameroun

Directeur
Département de la faune et des parcs nationaux
Ministère du tourisme
YAOUNDE
Cameroun

Tel: (237) 224411/222137
Tlx: 8318

Canada

Director
Canadian Parks Service
Environment Canada
Les Terrasses de la Chaudière
OTTAWA
Ontario
Canada KIA 1G2

Tel: (1) 819 994 1871
Tlx: 053 3608 parcs
FAX: (1) 819 997 2443

Executive Manager
Alberta Recreation and Parks
Standard Life Centre
10405 Jasper Avenue
EDMONTON
Alberta
Canada T5J 3N4

Tel: (1) 403 427 3948

Director
Parks Program Branch
Ministry of Environment & Parks
4000 Seymour Place
VICTORIA
British Columbia
Canada V8V 1X5

Tel: (1) 604 387 4317

Director
Parks Branch
Department of Natural Resources
258 Portage Street, 4th Floor
WINNIPEG
Manitoba
Canada R3C 1K2

Tel: (1) 204 945 4362

Executive Director
Field Operations
Tourism, Recreation & Heritage
P O Box 12345
FREDERICTON — Tel: (1) 506 453 2550
New Brunswick
Canada E3B 5C3

Director of Parks Division
Department of Environment and Lands
P O Box 8700
ST JOHN'S
Newfoundland
Canada AlB 4J6

Director
Parks Division
Department of Tourism and Parks
P O Box 2000
CHARLOTTETOWN — Tel: (1) 902 368 4275
Prince Edward Island
Canada ClA 7N8

Superintendent of Parks
Department of Economic Development
and Tourism
Government of the Northwest Territories
Box 1320
YELLOWKNIFE — Tel: (1) 403 873 7903
Northwest Territories
Canada XlA 2L9

Director
Parks and Recreation Division
Department of Lands and Forests
R.R. No. 1 Belmont
Colchester County — Tel: (1) 902 662 3030
Nova Scotia
Canada BOM 1CO

Parks and Recreation Areas Branch
Ministry of Natural Resources
Whitney Block
Queens Park
TORONTO — Tel: (1) 416 965 5160
Ontario
Canada M7A 1V3

Directeur de l'aménagement
Direction générale des parcs et des territoires fauniques
Ministère du loisir, de la chasse et de la pêche
150 boulevard St—Cyrille, 7e étage — Tel: (1) 418 643 5747
QUEBEC
Canada GIR 4Y1

Canadian Parks Service
220—10 Wellington Street
HULL — Tel: (1) 819 997 4931/994 2657
Quebec K1A 0H3 — FAX: (1) 819 994 5140
Canada

Director
Parks Branch
Parks, Recreation & Culture
532—3211 Albert Street
REGINA — Tel: (1) 306 787 2854
Saskatchewan
Canada S4S 5W6

Chief
Parks and Outdoor Recreation
Department of Renewable Resources
Government of the Yukon
Box 2703
WHITEHORSE
Yukon Territory
Canada YIA 2C6

Tel: (1) 403 667 5811/5802
Tlx: 368260

Cape Verde Islands

Divisão de Botanica
Instituto Nacional de Investigação Agraria
C P 84
Praia, SANTIAGO
Cape Verde

Cayman Islands

Senior Scientific Officer
Natural Resources Laboratory
PO Box 486
GRAND CAYMAN
Cayman Islands

Tel: 92557
Tlx: 4260 Cayman Gov. CP
FAX: (1) 809 949 7544

Central African Republic

Ministère des eaux, des forêts, chasses,
pêches et du tourisme
Direction générale du centre national pour la
protection et l'aménagement de la faune
B P 981
BANGUI
Central African Republic

Chad

Directeur
Direction des Parcs nationaux et Réserves de faune
B P 905 — Tel: 512305
NDJAMENA — FAX: 514397
Chad

Chile

Jefe
Departamento Areas Silvestres Protegidas
Corporación Nacional Forestal
Avda. Bulnes 259 Of. 604 — Tel: (56) 2 696 0783/1257
SANTIAGO — Tlx: 240001 conaf
Chile — FAX: (56) 2 671 5881

China

Director
National Parks Department
Ministry of Forestry
Hepingli
Dongcheng District
BEIJING — Tel: (86) 1 463061
People's Republic of China

Division of Nature Conservation
National Environmental Protection Agency
Ministry of Urban and Rural Construction
and Environmental Protection
Baiwanzhuang — Tel: (86) 1 899 2211
BEIJING — Tlx: 22477
People's Republic of China

Colombia

Jefe
Division de Parques Nacionales
Instituto Nacional de los Recursos
Naturales Renovables y del Ambiente
Apartado Aéreo 13458
BOGOTA DE
Colombia

Tel: (57) 1 283 2598/283 0964
Tlx: 44428
FAX: (57) 1 286 8643

Comoros

Direction générale de l'environnement,
de l'urbanisme et de l'habitat
Ministère de l'Equipement et de l'Environnement
BP 12
MORONI
Comoros

Congo

Directeur
Direction de la Conservation de la Faune
Ministère de l'économie forestière
B P 2153
BRAZZAVILLE, Congo

Tel: (242) 811718/813718

Cook Islands

Director
Conservation Service
Ministry of Internal Affairs & Conservation
P O Box 781
Tupapa, RAROTONGA
Cook Islands

Tel: (682) 20959
Tlx: 62056 RG
FAX: (682) 22950

Costa Rica

Director
Servicio de Parques Nacionales
Ministerio de Recursos Naturales,
Energia y Minas
Apartado 10094 Tel: (506) 335673/336213
1000 SAN JOSE FAX: (506) 338840
Costa Rica

Côte d'Ivoire

Directeur
Direction de la protection de la nature, de la
pisciculture, de l'aquaculture, des pêches en
eaux continentales et de l'environnement hydrographique
Ministère des eaux et forêts
B P V178
ABIDJAN
Côte d'Ivoire

Cuba

Comisión Nacional para la Protección del
Medio Ambiente y Conservación de los
Recursos Naturales (COMARNA)
Ave. 17 #5008 entre 50 y 52
Playa Tel: (53) 729 0501
C. DE LA HABANA FAX: (53) 762 5604
Cuba

Cyprus

Ministry of Agriculture &
Natural Resources
Nature Conservation Service Tel: (357) 2 402586
NICOSIA Tlx: 4660
Cyprus FAX: (357) 2 445156

Director
Department of Forests
Ministry of Agriculture & Forestry
Lefkosa—Nicosia — Tel: (90) 741 207 1315/1172
Via MERSIN 10 — FAX: *via* (90) 520 81111
Turkish Republic of Northern Cyprus

Czechoslovakia

Statni Ustav Pamatkove Pece a Ochrany Prirody
Valdstejnske Namesti 1
118 01 PRAHA 1 - MALA STRANA — Tel: (42) 2 513
Czechoslovakia

Ustredie statnej ochrany prirody
Liptovsky Mikulas
Stredisko rozvoja ochrany prirody
Heyrovsteho 1
84103 BRATISLAVA
Slovenia
Czechoslovakia

Denmark

Senior Advisor
National Forest & Nature Agency
Ministry of the Environment
Slotsmarken 13 — Tel: (45) 765376
2970 HORSHOLM — Tlx: 21485
Denmark — FAX: (45) 765477

Greenland Home Rule
Direktoratet for Boliger, Teknik og Miljo
Dept. Fysisk Planlaegning & Naturforvaltning
Postbos 1070
3900 NUUK
Greenland

Djibouti

Service de l'agriculture, de l'élevage,
des eaux et des forêts
Ministère de l'agriculture et du développement rural
B P 224
DJIBOUTI

Dominica

Chief Forester
National Park Service
Forestry Office
Ministry of Agriculture
P O Box 71
Botanical Gardens
ROSEAU
Dominica

Tel: (1) 809 445 2732
FAX: (1) 809 448 4815/2401

Dominican Republic

Director
Dirección Nacional de Parques
Apartado 2487
SANTO DOMINGO
Republica Dominicana

Tel: (1) 809 685 1316/
809 682 7628
Tlx: 6101 AGEMIR DA
FAX: (1) 809 685 3366

Ecuador

Director
Natural Areas & Wildlife Division
Direccion Nacional Forestal
Ministerio de Agricultura
Avdas Eloy Alfaro y Amazonas
QUITO
Ecuador

Tel: (593) 2 548924
/541988/541955

Egypt

Egyptian Environmental Affairs Agency
11(A) Hassan Sabry Street
Zamalek
CAIRO
Egypt

Tel: (20) 2 341 6566
FAX: (20) 2 342 0768

Egyptian Wildlife Service
Giza Zoo
12613 GIZA
Egypt

El Salvador

Jefe
Servicio de Parques Nacionales y
Vida Silvestre
Centro de Recursos Naturales (CENREN)
Ministerio de Agricultura y Ganaderia
Canton El Matazano
Aptdo 2265
SOYAPANGO
El Salvador

Tel: (503) 270622/235385
FAX: (503) 242478

Equatorial Guinea

Vice Ministro Delegado
Ministerie Delegado de Presidencia
encargado de aguas, bosques y forestal
MALABO
Equatorial Guinea

Tel: (240) 2119

Estonia

Ministry of Environment
Nature Conservation Department
TORTU
Estonia

Tel: (7) 14 343 4880/3996
Tlx: 173245 city su c/o Linnavolikogu
FAX: (7) 14 343 1466 c/o Linnavolikogu

Ministry of Environment
Toompaiestee 24 — Tel: (7) 14 245 2693
200 100 TALLINN — Tlx: 173238 mets su
Estonia — FAX: (7) 14 245 3310

Ethiopia

Manager
Ethiopian Wildlife Conservation Development and Organisation
Ministry of Agriculture, Nature Conservation
& Development
P O Box 386 — Tel: (251) 1 57532/
ADDIS ABABA — Tlx: 21036
Ethiopia — FAX: (251) 1 518977

Federated States of Micronesia

Secretary
Department of Human Resources
P O Box 490 — Tel: (691) 320 2619
KOLONIA, Pohnpei — Tlx: 6807 FSMGOV FM
Eastern Caroline Islands, FSM USA 96941 — FAX: (691) 379 2823

Fiji

Conservator of Forests
Ministry of Forests
P O Box 2218
SUVA — Tel: (679) 314591
Fiji

Environmental Planner
Department of Town & Country Planning
Ministry of Housing & Urban Affairs
P O Box 2350 — Tel: (679) 211 208
SUVA — Tlx: 2167 FOSEC FJ
Fiji — FAX: (679) 303 515

Director
National Trust for Fiji
P O Box 2089
Government Buildings
SUVA
Fiji

Tel: (679) 301807
FAX: (679) 302646

Finland

Chief
Office for National Parks
National Board of Forestry
P O Box 94
01301 VANTAA
Finland

Tel: (358) 0 857 841
FAX: (358) 0 857 84200

France

Directeur
Protection de la Nature
Ministère de l'Environnement et Cadre de la Vie
Direction de la protection de la nature
14 boulevard du Général Leclerc
92524 NEUILLY—SUR—SEINE
France

Tel: (33) 14 758 1212
Tlx: 620602 DENVIR F
FAX: (33) 14 745 0474/2360

(Antarctic – Ile Amsterdam, Saint Paul, Crozet, Kerguelen)
Administrateur Supérieur
Terres Australes et Antarctiques Françaises
34 rue des Renaudes
75017 PARIS
France

French Guiana

Direction départementale de l'agriculture
CAYENNE
French Guiana

French Polynesia

Spécialiste de la gestion des périmètres protégés
Délégation à l'environnement, Ministère de l'environnement
BP 4562
PAPEETE — Tel: (689) 432409
Tahiti — Tlx: GVPF 404 FP
French Polynesia — FAX: (689) 419252

Gabon

Directeur de la faune et de la chasse
Ministère des eaux et forêts
B P 1128 — Tel: (241) 748561
LIBREVILLE
Gabon

Gambia

Wildlife Conservation Department
c/o Ministry of Water Resources, Forestry & Fisheries
5 Marina Parade — Tel: (220) 27307
BANJUL — Tlx: 2204
Gambia

Germany

Bundesforschungsanstalt für Naturschutz und Landschaftsökologie
Konstantinstrasse 110
5300 BONN 2 — Tel: (49) 228 849 1190
Germany (W) — FAX: (49) 228 849 1200

Ghana

Chief Game and Wildlife Officer
Ministry of Lands & Natural Resources
Department of Game and Wildlife
P O Box M 239 — Tel: (233) 21 666129
ACCRA
Ghana

Ghana Forestry Commission
P O Box M239 — Tel: (233) 21 220818/776145
ACCRA
Ghana

Gibraltar

Chief Minister
Government of Gibraltar
GIBRALTAR

Greece

Head
Section for the Forest Environment, National
Parks and Forest Recreation
Directorate of Aesthetic Forests, Parks & Game Management
Ministry of Agriculture — Tel: (30) 1 363 7659
3—5 Ippocratous Street — FAX: (30) 1 360 7138
101 64 ATHENS, Greece

Grenada

National Parks & Wildlife Unit
Forestry Department
Ministry of Agriculture
ST GEORGE'S — Tel: (1) 809 440 3083/2934
Grenada

Guadeloupe

Office national des Forêts
BP 648
Jardin botanique
97109 BASSE—TERRE — Tel: (590) 811720
Guadeloupe — FAX: (590) 814877

Guatemala

Jefe
Departamento de Vida Silvestre y
Areas Protegidas (DIGEBOS)
Ministerio de Agricultura, Ganadería y
Alimentación
7a Avda 6—80 Zona 13
CIUDAD DE GUATEMALA
Guatemala

Tel: (502) 2 720509

Secretario Ejecutivo
Consejo Nacional de Areas Protegidas
Presidencia de la Republica
7a Av. 4—00, Zona 1
CIUDAD DE GUATEMALA
Guatemala

Tel: (502) 2 21816/532477
FAX: (502) 2 535109

Guinea

Coordonnateur National
Division Protection de la Nature et
des Ressources naturelles
Direction nationale de l'Environnement
Ministère des Ressources naturelles
et de l'Environnement
BP 295
CONAKRY
Guinea

Tel: (224) 22350

Chef de la Section
Protection de la Nature
Direction nationale des eaux
et forêts
BP 624
CONAKRY
Guinea

Tel: (224) 443249
Tlx: 22338 mara
FAX: (224) 444387

Guinea—Bissau

Director Geral
Direcção Geral dos Serviços Florestais e Caça
Ministerio do Desenvolvimento Rural e Agricultura
C P 71 Tel: (245) 215475/213176
BISSAU
Guinea—Bissau

Guyana

Executive Chairman
Guyana Agency for Health Sciences, Education,
Environment and Food Policy (GAHEF)
Liliendaal Tel: (592) 2 575234
GREATER GEORGETOWN
Guyana

Haiti

Institut de sauvegarde du patrimoine
national (ISPAN)
B P 2484
PORT—AU—PRINCE
Haiti

Ministère de l'agriculture,
des ressources naturelles et du développement rural
Direction des ressources naturelles
BP 1441
PORT—AU—PRINCE Tel: (509) 1 21862
Haiti

Honduras

Jefe
Depto de Vida Silvestre y Recursos
Ambientales
Ministerio de Recursos Naturales Renovables
Blvd Miraflores
Apartado 309
TEGUCIGALPA
Honduras

Tel: (504) 326213/326227
Tlx: 8071 Serena

Hong Kong

Director
Agriculture and Fisheries Department
Canton Road Government Offices
393 Canton Road, 12th floor
Kowloon
HONG KONG

Tel: (852) 3 733 2136
FAX: (852) 3 311 3731

Hungary

Környezet—védelmi és Területfejlesztési Minisztérium
Orszagos Termeszetredelmi Hitaval
Költö u. 21
PO Box 351
1394 BUDAPEST
Hungary

Tel: (36) 1 562133
Tlx: 226115
FAX: (36) 1 757457

Iceland

Director
Nature Conservation Council
Hverfisgata 26
101 REYKJAVIK
Iceland

Tel: (354) 1 22520

India

Joint Secretary (Wildlife)
Department of Environment, Forests and Wildlife
Ministry of Environment and Forests
Paryavaran Bhawan
CGO Complex, Lodi Road — Tel: (91) 11 306156/362281
NEW DELHI 110 003 — Tlx: 66185
India

Regional Deputy Director (Wildlife)
Eastern Region
Nizam Palace
A J C Bose Road
CALCUTTA 700 020
India

Regional Deputy Director (Wildlife)
Western Region
Air Cargo Complex
Sahaw
BOMBAY 400 099
India

Regional Deputy Director (Wildlife)
Southern Region
2/C/S Brownstone Appts.
Mahalingapuram
MADRAS 600 034
India

Regional Deputy Director (Wildlife)
Northern Region
Bikaner House Barracks
Shahjahan Road
NEW DELHI 110 019
India

Chief Wildlife Warden
P O Chatham
PORT BLAIR
Andaman & Nicobar 744 101, Union Territory
India

Conservator of Forests (Wildlife)
HYDERABAD
Andhra Pradesh 500 004
India

Chief Wildlife Warden
ITANAGAR
Arunachal Pradesh 791 111
India

Chief Conservator of Forests &
Chief Wildlife Warden
Retabari
GUWAHATI
Assam 788 735
India

Chief Wildlife Warden
P O Hinoo
RANCHI
Bihar 834 002
India

Divisional Forest Officer &
Chief Wildlife Warden
Estate Office, Sector 17
CHANDIGARH 160 017, Union Territory
India

Chief Wildlife Warden
SILVASSA
Dadra & Nagar Haveli 396 230
Union Territory
India

ADM and Chief Wildlife Warden
Room 39, 1st Floor
Western Wing
Tis Hazari
DELHI 110 054
Union Territory
India

Conservator of Forests & Chief
Wildlife Warden
Wildlife & Parks Division
Junta House, 3rd Floor
PANAJI
Goa 403 001
India

Chief Conservator of Forests &
Chief Wildlife Warden
Kothi Annexe
BARODA
Gujarat 390 001
India

Chief Wildlife Warden
Kothi 70, Sector 8
PANCHKULA
Haryana 134 109
India

Chief Conservator of Forests (Wildlife)
& Chief Wildlife Warden
Talland
SHIMLA
Himachal Pradesh 171 002
India

Chief Wildlife Warden
Tourist Reception Centre
SRINAGAR
Jammu & Kashmir 190 001
India

Chief Wildlife Warden
11th Floor
Aranya Bhawan, 18th Cross
Malleshwaram
BANGALORE
Karnataka 560 003
India

Additional Chief Conservator of
Forests & Chief Wildlife Warden
TRIVANDRUM
Kerala 695 001
India

Administrator
Via KAVARATTI
Lakshadweep 673 555
Union Territory
India

Chief Conservator of Forests (Wildlife)
& Chief Wildlife Warden
M.P. Forest Dept.
1st Floor, B—Wing
Satpura Bhawan
BHOPAL
Madhya Pradesh 462 001
India

Chief Wildlife Warden
Nature Conservation MS
NAGPUR
Maharashtra 440 001
India

Chief Wildlife Warden
Sanjenthong P O
IMPHAL
Manipur 795 001
India

Chief Wildlife Warden
Risa Colony
SHILLONG
Meghalaya 793 003
India

Chief Wildlife Warden
AIZAWL
Mizoram 796 001
India

Chief Conservator of Forests (Wildlife)
DIMAPUR
Nagaland 797 112
India

Additional Conservator of Forests (Wildlife)
& Chief Wildlife Warden
315 Kharvel Nagar
BHUBANESHWAR
Orissa 751 001
India

Chief Wildlife Warden
PONDICHERRY 605 001
Pondicherry
Union Territory
India

Chief Wildlife Warden
Punjab SCD 2463—64, Sector 22C
CHANDIGARH
Punjab 160 022
India

Chief Wildlife Warden
Van Bhawan, Bhagwan Das Road
JAIPUR
Rajasthan 302 001
India

Conservator of Forests (Fish
& Wildlife) & Chief Wildlife Warden
GANGTOK
Sikkim 737 101
India

Additional Conservator of Forests (Wildlife)
& Chief Wildlife Warden
571 Tiruchi Road
COIMBATORE
Tamil Nadu 641 045, India

Chief Conservator of Forests & Chief
Wildlife Warden
P O Kunjaban
AGARTALA
Tripura 799 006
India

Chief Wildlife Warden
17 Rana Pratap Marg
LUCKNOW
Uttar Pradesh 226 001
India

Chief Conservator of Forests &
Chief Wildlife Warden
P—16, India Exchange Place Ext.
New Cit Bldg.
CALCUTTA
West Bengal 700 073
India

Indonesia

Directorate of National Parks & Recreation Forests
Jalan Ir H Juanda 15
PO Box 133
BOGOR 16001
Indonesia
Tel: (62) 251 321014/323972
FAX: (62) 251 584818

Director
Directorate General of Forest Protection
and Nature Conservation
Jalan Ir H Juanda 15
BOGOR 16122
Indonesia
Tel: (62) 251 324013/323067
Tlx: 45996

Head
National Parks Division
Ministry of Forestry
Manggala Wanabakti Building
Jl Gatot Subroto
JAKARTA
Indonesia

Tel: (62) 21 583033
FAX: (62) 21 570026
Tlx: 45996 dephut ia

Iran

Director
Department of the Environment
P O Box 15875
5181 TEHRAN
Iran

Tel: (98) 21 966441/
891261
Tlx: 215064

Iraq

Higher Council for Environmental Protection
and Improvement
c/o University of Health
P O Box 423
BAGHDAD
Iraq

Ireland

National Parks and Wildlife Service
Office of Public Works
51 St Stephen's Green
DUBLIN 2
Republic of Ireland

Tel: (353) 1 613111
Tlx: 90160 opw ei
FAX: (353) 1 610747

Israel

Director General
National Park Authority
P O Box 7028
HAKIRA
Israel 61070

Tel: (972) 3 252281/267651
FAX: (972) 3 267643

Director General
Nature Reserves Authority
78 Yirmiyahu St
JERUSALEM 94467 Tel: (972) 2 387461
Israel

Italy

Direzione General
Servizio Conservazione Natura
Ministerio dell'Ambiente
Via Volturno 58 Tel: (39) 6 445 1317
00187 ROME FAX: (39) 6 445 2831
Italy

Jamaica

Natural Resources Conservation Department
National Parks Branch
P O Box 305
KINGSTON 10 Tel: (1) 809 923 5155
Jamaica (1) 809 923 5070/5166

Japan

Director
Wildlife Protection Division
Nature Conservation Bureau
Environment Agency
1—2—2 Kasumigaseki
Chiyoda—Ku Tel: (81) 33 581 3351
TOKYO Tlx: 33855 jpnea j
Japan 100 FAX: (81) 33 595 1715/6

Jordan

Director
National Parks & Historic Monuments Department
Ministry of Tourism
P O Box 224 — Tel: (962) 6 642311
AMMAN — FAX: (962) 6 648465
Jordan

The Royal Society for the Conservation of Nature
P O Box 6354 — Tel: (962) 6 811689
AMMAN — Tlx: 21456
Jordan — FAX: (962) 6 628258

Kenya

Kenya Wildlife Service
Ministry of Tourism & Wildlife
P O Box 40241 — Tel: (254) 2 501081/2
NAIROBI — Tlx: 25016
Kenya — FAX: (254) 2 505866

Kiribati

Director
Wildlife Conservation Unit
Ministry of the Line and Phoenix Islands
KIRITIMATI (Christmas Island), Kiribati

Korea

Department of Forest Management
PYONGYANG
Democratic People's Republic of Korea

National Parks Authority
Gun—Sol—Kwi—Kwan 71—2
Non Hyung—dong Kangnam—gu
SEOUL 135 010 — Tel: (82) 2 720 5421
Republic of Korea

Kuwait

Environment Protection Council
P O Box 24885
13104 SAFAT — Tel: (965) 456833/452790
Kuwait — Tlx: 46408

Laos

Forest Department
Wildlife Conservation & Fisheries
Ministry of Industry, Handicrafts & Forests
VIENTIANE
Lao People's Democratic Republic

Latvia

Environmental Protection Committee
RIGA — FAX: (7) 13 222 8159
Latvia

Lebanon

Directeur des Forêts et des Ressources
Naturelles
Ministère de l'Agriculture
rue Sami Solh
BEIRUT
Lebanon

Lesotho

Lesotho National Parks
Conservation Division
Ministry of Agriculture and Marketing
P O Box 92
MASERU
Lesotho

Liberia

Managing Director
Division of Wildlife and National Parks
Forestry Development Authority
B P 3010
MONROVIA
Liberia

Tel: (231) 271865
Tlx: c/o GFM 44230

Libya

Director General
Secretariat of Agriculture and Agrarian Reform
Forests Section
P O Box 12901
TRIPOLI
Libya

Lithuania

Head
Foreign Relations Division
Environmental Protection Division
Juozapaviciaus Str. 9
232 600 VILNIUS
Lithuania

Tel: (7) 12 235 8275
Tlx: 261191 gamta su
FAX: (7) 12 235 8020

Liechtenstein

Geschäftsstelle
Gesellschaft für Umweltschutz
Heiligkreuz 52
9490 VADUZ — Tel: (41 75) 25262
Liechtenstein

Luxembourg

Chef
Direction des eaux et forêts
Service de conservation de la nature
B P 411
34 avenue de la Porte—Neuve — Tel: (352) 402201
2227 LUXEMBOURG
Luxembourg

Macao

Servicios Florestais e Agricolas
Rua Central 107
MACAO

Madagascar

Chef
Service de la protection de la nature
Direction des eaux et forêts
Ministère de la production animale (élévage et pêche),
des eaux et des forêts
B P 243, Nanisana — Tel: (261) 2 40811
101 ANTANANARIVO — Tlx: 22520 mpaef mg
Madagascar

Malawi

Chief Parks and Wildlife Officer
Department of National Parks & Wildlife
P O Box 30131
LILONGWE 3
Malawi
Tel: (265) 730944
Tlx: 4113

Malaysia

Director—General
Department of Wildlife & National Parks
Km 10 Jalan Cheras
50664 KUALA LUMPUR
Malaysia
Tel: (60) 3 905 2872/3/5
FAX: (60) 3 905 2873

Director—General
Sabah Parks
Box 10626
880 26 KOTA KINABALU
Sabah
Malaysia

Sabah Wildlife Department
Tinqkat 7 Kenara Saba Bank Bldg
Wisma Tun Fuad Stephens
88300 KOTA KINABALU
Sabah
Malaysia
Tel: (60) 88 52666
FAX: (60) 88 222476

Head
National Park & Wildlife Office
Jalan Stadium
Petra Jaya
93660 KUCHING
Sarawak
Malaysia
Tel: (60) 82 442180
FAX: (60) 82 441377

Mali

Direction nationale des eaux et forêts
Ministère de l'environnement et de l'élevage
B P 275
BAMAKO — Tel: (223) 225850
Mali

Malta

Environment Officer (Conservation)
Environment Division — Tel: (356) 235495/ 234848
Ministry of Education
BELTISSEBH — Tlx: 1114
Malta — FAX: (356) 222256

Marshall Islands

General Manager
Environmental Protection Agency
P O Box 1322
MAJURO
Marshall Islands 96960

Martinique

Directeur
Parc Naturel Regional
Ancienne Collège Agricole de Tivoli
B P 437 — Tel: (596) 64 42 59
97205 FORT—DE—FRANCE Cédex — FAX: (596) 64 72 27
Martinique

Mauritania

Direction de la protection de la nature
Ministère du développement rural
B P 170 — Tel: (222) 251836
NOUAKCHOTT — Tlx: 585
Mauritania — FAX: (222) 251834

Mauritius

Conservator of Forests
Forestry Service
Botanical Garden Street
Les Casernes
CUREPIPE
Mauritius

Conservation Unit
Ministry of Agriculture, Fisheries
and Natural Resources
PORT LOUIS
Mauritius

Mexico

Director General
Dirección General de Parques
SEDUE
Avda Constituyentes 947
Edificio B, PB
Colón Belen de las Flores
CP 01100
MEXICO DF
Mexico
Tel: (52) 5 271 2640/1270
Tlx: 1771198 Sedu Me
FAX: (52) 5 271 6614

Director General
Dirección General de Conservación Ecológica
de los Recursos Naturales
Secretaria de Desarrollo Urbano y Ecología
Rio Elba No. 20 – piso 10°
Col. Cuauhtemoc
06500 MEXICO DF
Mexico
Tel: (52) 5 286 9278/9276
FAX: (52) 5 286 6625

Monaco

Association Monagasque pour la conservation
de la nature
7 rue de Colle
MONACO MC 98000 — Tel: (33 93) 302107
Principauté de Monaco

Mongolia

Central Council
Mongolian Association for Conservation of Nature
and Environment
CPO Box 530 — Tel: 26330
ULAN BATOR 12
Mongolia

Montserrat

Coordinator, National Parks
Montserrat National Trust
Box 332
Plymouth — Tel: (1) 809 491 2687
MONTSERRAT, West Indies

Morocco

Chef
Division de la chasse, de la pêche et de la protection
de la nature
Administration des eaux et forêts
Ministère de l'agriculture et de la réforme agraire
1 rue Jaffaar Essadig — Tel: (212) 7 62694/63946
RABAT Chellah — FAX: (212) 7 764446
Morocco

Mozambique

Director
Servico de Conservação de Fauna Bravia
Departamento de Florestas e Fauna Bravia
Ministeiro de Agricultura
Caixa Postal 3651
MAPUTO
Mozambique

Myanmar (Burma)

Director
Wildlife & Sanctuaries Division
Forest Department
Gyogon, Insein — Tel: (95) 1 63490
YANGON (Rangoon)
Union of Myanmar (Burma)

Namibia

Directorate of Nature Conservation
and Recreation Resorts
Private Bag 13306 — Tel: (264) 61 63131
WINDHOEK 9000 — Tlx: 908 3180
Namibia — FAX: (264) 61 63195

Nepal

Director General
Department of National Parks and Wildlife Conservation
P O Box 860
Babar Mahal — Tel: (977) 1 220912/227926/220850
KATHMANDU — Tlx: 2567 kmtnc np
Nepal — FAX: (977) 1 410073/26820

Secretary
King Mahendra Trust for Nature Conservation
P O Box 3712
Babar Mahal — Tel: (977) 1 526573/526571
KATHMANDU — Tlx: 2567 kmtnc np
Nepal — FAX: (977) 1 526570

Netherlands

Director
Department of Nature Conservation
Ministry of Agriculture, Nature Management & Fisheries
Postbus 20401 — Tel: (31) 70 379 3000
2500 EK 's GRAVENHAGE — Tlx: 32040
The Netherlands — FAX: (31) 70 347 8228

Netherlands Antilles

Stichting Nationale Parken Nederlandse
Antillean/CARMABI
P O Box 2090
WILLEMSTAD — Tel: (599) 9 624242/624705
Curaçao — FAX: (599) 9 627680
Netherlands Antilles

Stichting Nationale Parken Nederlandse
Antillean
P O Box 368
KRALENDIJK
Bonaire
Netherlands Antilles

Saba Conservation Foundation
c/o Administration Building
The Bottom — Tel: (599) 4 63348
SABA — FAX: (599) 4 63348
Netherlands Antilles

New Caledonia

Service de l'Environnement et Gestion des
Parcs et Réserves
Direction du Développement Rural
B P 256
NOUMEA CEDEX
New Caledonia

New Zealand

Director General
Department of Conservation
P O Box 10420
WELLINGTON
New Zealand
Tel: (64) 4 471 0726
FAX: (64) 4 473 3656

Nicaragua

Director
Servicio de Parques Nacionales
Dirección General de Recursos Naturales
y del Ambiente
Apartado 5123
MANAGUA
Nicaragua
Tel: (505) 2 31110/13
Tlx: 1328
FAX: (505) 2 31274

Niger

Ministère de l'agriculture et de l'environnement
Direction de la faune, de la pêche et de la pisciculture
B P 721
NIAMEY
Niger
Tel: (277) 734069
Tlx: 5386

Nigeria

Director of Forestry
Federal Ministry of Agriculture and Water Resources
P M B 12613
6 Ijeh Village Road
Obalande
Ikoyi Tel: (234) 1 684178
LAGOS
Nigeria

North Mariana Islands

Division of Parks & Recreation
Department of Natural Resources
SAIPAN
MP 96950
USA

Norway

Head of Division
Nature Conservation Division
Royal Ministry of Environment
P O Box 8013 Tel: (47) 2 349090/419010
0030 OSLO 1 Tlx: 21480 env n
Norway FAX: (47) 2 349560

Directorate for Nature Management
Tungasletta 2 Tel: (47) 7 580500
7004 TRONDHEIM FAX: (47) 7 915433
Norway

(Antarctic - Bouvetoya)
Director
Justis—og politidepartement
Polaravdelingen
P O Box 8005 Dep. Tel: (47) 2 419010
0030 OSLO 1
Norway

Oman

Adviser for Conservation of the Environment
Office of the Adviser for Conservation
of the Environment
Diwan of Royal Court Affairs
P O Box 246
The Palace Tel: (968) 722482/333
MUSCAT Tlx: 5667
Sultanate of Oman FAX: (968) 740550

Ministry of Environment and Water Resources
P O Box 323 Tel: (968) 696458
MUSCAT Tlx: 5405
Oman FAX: (968) 602320

Pakistan

Conservator (Wildlife)
National Council for Conservation of Wildlife
Ministry of Food, Agriculture and Cooperatives
485 Street 84, G—6/4
ISLAMABAD Tel: (92) 51 823520
Pakistan Tlx: 5844 minfa

Wildlife Warden
Wildlife Wing
Forest Department — Tel: 18
MOZAFFARABAD
Azad State of Jammu & Kashmir
Pakistan

Divisional Forest Officer (Wildlife)
Forestry & Wildlife Department
Government of Baluchistan — Tel: (92) 81 71298
Spinny Road
QUETTA
Baluchistan
Pakistan

Director
Environment Directorate
Capital Development Authority
Sitara Market — Tel: (92) 51 826397
ISLAMABAD FCT
Pakistan

Conservator of Forests
P O Box 501
GILGIT — Tel: 360
Northern Areas
Pakistan

Conservator (Wildlife)
Forest Department (Wildlife Wing)
Government of NWFP Forest Office
Shami Road
PESHAWAR — Tel: (92) 521 73184
North—West Frontier Province — FAX: (92) 521 61235
Pakistan

Conservator of Forests (Parks & Wildlife)
Wildlife Department
Government of Punjab
2 Sanda Road
LAHORE
Punjab
Pakistan

Tel: (92) 42 61798/63947

Conservator of Forests (Wildlife)
Sind Wildlife Management Board
Aiwan—e—Saddar Road
P O Box 3722
Stratchen Road
KARACHI 1
Sind
Pakistan

Tel: (92) 21 523176
Cable: WILDLIFE

Palau

Chief Conservationist
Division of Conservation & Entomology
Biology Laboratory
P O Box 100
KOROR
Republic of Palau

Tel: (680) 490/408
Tlx: 8914

Panama

Director Nacional
Areas protegidas y Vida Silvestre
Instituto Nacional de Recursos
Naturales Renovables (INRENARE)
Paraiso, Ancon
Apartado 2016
CIUDAD DE PANAMA
Panama

Tel: (507) 324518

Papua New Guinea

Secretary
Department of Environment and Conservation
P O Box 6601
BOROKO NCD
Papua New Guinea

Tel: (675) 271788
Tlx: 22327 NE
FAX: (675) 271044

Paraguay

Director
Dirección de Parques Nacionales y Vida Silvestre
25 de Mayo 640 c/Antequera
Edificio Garantia Piso 12A
ASUNCION
Paraguay

Tel: (595) 21 494914
FAX: (595) 21 495568

Peru

Director de Conservación
Instituto Nacional Forestal y de Fauna (INFOR)
Jr Cahuide 805, 7° piso
Jesus Maria
LIMA 11
Peru

Director de Parques Nacionales
Dirección General de Forestal y Fauna
Ministerio de Agricultura
Natalio Sanchez 220 of. 907
Apartado 11—0150
Jesus Maria
LIMA 11
Peru

Tel: (51) 14 323150
Tlx: 20053
FAX: (51) 14 232789

Philippines

Protected Areas & Wildlife Bureau
Department of Environment and Natural Resources
Quezon Blvd
Diliman
QUEZON CITY
Philippines

Tel: (63) 2 978511—15
Tlx: 7572000 envinar ph
FAX: (63) 2 981010

Pitcairn

Office of the Governor of Pitcairn,
Henderson, Ducie & Oeno Is.
c/o British High Commission
P O Box 1812
WELLINGTON
New Zealand

Poland

Ministerstwo Ochrony Srodowiska
i Zasobow Naturalnych
ul. Wawelska 52/54
00922 WARSAW
Poland

Tel: (48) 2 225 0001/226204
Tlx: 817157
FAX: (48) 2 391 20049

Portugal

Presidente
Serviço Nacional de Parques, Reservas e
Conservaçao de Natureza
Rua da Lapa 73
1200 LISBON
Portugal

Tel: (351) 1 675259/675395
Tlx: 44089
FAX: (351) 1 601048

Puerto Rico

Department of Natural Resources
Box 5887 — Tel: (1) 809 723 1464/3090
Puerta De Terra — FAX: (1) 809 724 0365/722 2785
00906 PUERTO RICO

Qatar

Ministry of Industry & Agriculture
Agricultural Affairs Administration
P O Box 1966 — Tel: (974) 433400
DOHA — Tlx: 4751
Qatar

Environment Protection Committee
c/o Ministry of Public Health
P O Box 42 — Tel: (974) 320825
DOHA — FAX: (974) 415246
Qatar

Réunion

Office national des forêts
Direction régionale de la Réunion
Colline de la Providence
97488 SAINT—DENIS, Réunion

Romania

Ministry of Environment
Bd Libertatu
Artera Noua N—5, Tronson 5—6
Sectorul 5 — Tel: (40) 0 316004
BUCURESTI 22 — Tlx: 11457
Romania — FAX: (40) 0 316199

Rwanda

Chef du Service des parcs nationaux, tourisme
et agence de voyages
Office Rwandais du tourisme et des parcs nationaux
B P 905
KIGALI
Rwanda

San Marino

Département des Affaires Etrangères
Contrada Omerelli
Palazzo Begui
Via Giacomini
SAN MARINO

Sao Tomé & Principe

Commisão de Coordenação Florestal
Ministerio de Agricultura e Pescas
CP 47
SAO TOME
Sao Tomé & Principe

Saudi Arabia

Directorate of National Parks
Ministry of Agriculture & Water
RIYADH 11195 — Tel: (966) 1 402 0268
Kingdom of Saudi Arabia

Director General
Natural Resources Environment Protection
Meteorology & Environment Protection Administration
P O Box 1358 — Tel: (966) 2 651 2312
JEDDAH 21431 — Tlx: 401236
Kingdom of Saudi Arabia — FAX: (966) 2 651 1124

National Commission for Wildlife Conservation
& Development
P O Box 61681 Tel: (966) 1 441 8700
RIYADH 11575 Tlx: 405930 sncwed sj
Kingdom of Saudi Arabia FAX: (966) 1 441 0797

Senegal

Directeur
Direction des parcs nationaux
Ministère de la protection de la nature
B P 5135
DAKAR FANN, Senegal Tlx: (221) 3248

Seychelles

Principal Secretary
Department of Environment
B P 445 Tel: (248) 24744
Victoria FAX: (248) 24500
MAHE
Republic of Seychelles

Principal Secretary
Seychelles Island Foundation
P O Box 445
Victoria Tel: (248) 25333
MAHE Tlx: 2312
Republic of Seychelles FAX: (248) 21787

Sierra Leone

Superintendent of Wildlife and National Parks
Wildlife Conservation Branch
Forestry Division
Ministry of Agriculture & Natural Resources
Tower Hill Tel: (232) 22 24821
FREETOWN
Sierra Leone

Singapore

Executive Director
National Parks Board
Singapore Botanic Gardens
Cluny Road
SINGAPORE 1025

Tel: (65) 474 1165/
470 9918
FAX: (65) 475 4295

Commissioner
Parks & Recreation Department
5th Storey
National Development Building Annex B
7 Maxwell Road
SINGAPORE 0106

Tel: (65) 322 6410/6416
Tlx: 22603 PRD
FAX: (65) 322 6422

Solomon Islands

Principal Conservation Officer
Environment Conservation Division
Ministry of Natural Resources
P O Box G24
HONIARA
Solomon Islands

Tel: (677) 23696
Tlx: 66306 SOLNAT HQ
FAX: (677) 21245

Somalia

Department of Wildlife
National Range Agency
Ministry of Livestock, Range and Forestry
B P 1759
MOGADISHU
Somalia

South Africa

Chief Director
National Parks Board
P O Box 787
PRETORIA 0001
Republic of South Africa

Tel: (27) 12 343 9770
Tlx: 321324
FAX: (27) 12 343 9959

The Director
Cape Department of Nature & Environmental Conservation
Private Bag X9086
CAPE TOWN 8000
South Africa

Chief Director
Natal Parks, Game & Fish Preservation Board
P O Box 662
PIETERMARITZBURG 3200
South Africa

Tel: (27) 331 471961
Tlx: 643481
FAX: (27) 331 471037

The Deputy Director
Nature Conservation (National States)
Department of Cooperation & Development
P O Box 384
PRETORIA 0001
South Africa

The Director
Orange Free State Nature Conservation Division
P O Box 507
BLOEMFONTEIN 9300
South Africa

The Director
Nature Conservation Division
Transvaal Provincial Administration
Private Bag X209
PRETORIA 0001
South Africa

The Director
National Parks Board
Private Bag X2078
MAFIKENG 8670
Republic of Bophuthatswana
South Africa

The Director—General
Department of Agriculture & Forestry
Private Bag X501
ZWELITSHA 5600
Republic of Ciskei
South Africa

The Director
Department of Agriculture & Forestry
Private Bag X5002
UMTATA
Republic of Transkei
South Africa

The Secretary
Department of Agriculture & Forestry
Private Bag X2247
SIBASA
Republic of Venda
South Africa

(Antarctic – Prince Edward Islands)
Director
Subdirectorate on Antarctica and Islands
Department of Environment Affairs
Private Bag X447
PRETORIA 0001
South Africa

Tel: (27) 12 310 3674
FAX: (27) 12 322 2682
Tlx: 320142

Spain

Sub Director General de Recursos Naturales Renovables
Instituto Nacional para la Conservación
de la Naturaleza (INCONA)
Gran Via de San Francisco — Tel: (34) 1 347 6159/6189
28005 MADRID — Tlx: 47591 aeico e
Spain — FAX: (34) 1 265 8379/347 6301

Sri Lanka

Director
Department of Wildlife Conservation
Ministry of Lands, Irrigation & Mahaweli Development
82 Rajamalwatte Road
Battaramulla — Tel: (94) 1 566601/567084
COLOMBO 6 — Cable: WILDLIFE
Sri Lanka — FAX: (94) 1 567088

Conservator of Forests
Forest Department
Rajamalwatte Road — Tel: (94) 1 566634
Battaramulla — FAX: (94) 1 580089
COLOMBO
Sri Lanka

St Christopher—Nevis

The Physical Planning Officer
Planning Unit
P O Box 186
BASSETERRE
St Christopher—Nevis
West Indies

St Helena

Director
Agriculture & Natural Resources
Government of St Helena
JAMESTOWN
St Helena
South Atlantic

St Lucia

Chief Forestry & Lands Officer
Department of Forestry
Ministry of Agriculture
CASTRIES — Tel: (1) 809 452 3078
St Lucia — FAX: (1) 809 453 6087
West Indies

Director
St Lucia National Trust
P O Box 525
CASTRIES — Tel: (1) 809 425 5005
St Lucia
West Indies

St Vincent & the Grenadines

The Director
Ministry of Agriculture
KINGSTOWN
St Vincent and the Grenadines

Sudan

Director
Wildlife Management and Parks
Regional Ministry of Wildlife
Conservation, Fisheries and Tourism
P O Box 77
JUBA
Sudan

Director
Wildlife & National Parks Forces
Ministry of Interior
P O Box 336
KHARTOUM Tel: (249) 11 76486
Sudan

Suriname

Head
Nature Conservation Division
Suriname Forest Service
Postbox 436
10 Cornelis Jongbawstraat Tel: (597) 479431/475845
PARAMARIBO Tlx: 364
Suriname FAX: (597) 410256

Stichting Natuurbehoud Suriname (STINASU)
P O Box 436
PARAMARIBO Tel: (597) 75845 x 343541
Suriname

Swaziland

Chief Executive Officer
Swaziland National Trust Commission
P O Box 100
LOBAMBA Tel: (268) 61151
Swaziland

Sweden

Director General
Statens Naturvardsverk — Tel: (46) 8 799 1000
171 85 SOLNA — Tlx: 11131
Sweden — FAX: (46) 8 292382

Switzerland

Office Fédéral de l'environnement,
des forêts et du paysage
Division Protection de la Nature
Neufeldstr. 5
C P 5662 — Tel: (41) 31 618081
3001 BERN — FAX: (41) 31 617827
Switzerland

Syria

Ministry of State for Environmental Affairs
P O Box 3754 — Tel: (963) 11 226600
DAMASCUS — Tlx: 411930
Syria

Director of Forests
Ministry of Agriculture
Al Jabri Street — Tel: (963) 11 213613/6
DAMASCUS
Syria

Taiwan

Director General
National Parks Department
Ministry of Interior
194, Section 3
Peihsin Road
Hsintien City
TAIPEI HSIEN, Taiwan RoC

(for reserves)
Vice Chairman
Council of Agriculture
Executive Yuan
37 Nanhai Road — Tel: (886) 2 312 4045
TAIPEI — FAX: (886) 2 312 5857
Taiwan 10728 RoC

Tanzania

Director
Tanzania National Parks Authority
P O Box 9192 — Tel: (255) 51 26052/25284
DAR—ES—SALAAM — Tlx: 41246
Tanzania

Tanzania National Parks
P O Box 3134 — Tel: (255) 573471
ARUSHA — FAX: (255) 573472
Tanzania

Wildlife Division
Ministry of Tourism, Natural Resources and Environment
P O Box 9372 — Tel: (255) 51 27271
DAR—ES—SALAAM
Tanzania

Ngorongoro Conservation Area Authority
Box 1
NGORONGORO
Arusha Region
Tanzania

Thailand

Director
National Parks Division
Royal Forest Department
Phaholyothin Road
Bangkhen
BANGKOK 10900
Thailand

Tel: (66) 2 579 4294/4842/0529
FAX: (66) 2 579 1562/8532/2791

Director
Wildlife Conservation Division
Royal Forest Department
Phaholyothin Road
Bangkhen
BANGKOK 10900
Thailand

Togo

Direction des parcs nationaux, ressources
faune et chasse (eaux et forêts)
Ministère de l'environnement et du tourisme
B P 355
LOME
Togo

Tokelau

Director of Agriculture & Fisheries
Office of Tokelau Affairs
P O Box 865
APIA
Western Samoa

Tlx: 222 TAWA SX

Tonga

Parks and Reserves Authority
Ministry of Lands, Survey and Natural Research
P O Box 5 Tel: (676) 23120/21210/21511
NUKU'ALOFA Tlx: 66269 PRIMO TS
Tonga FAX: (676) 23216

Trinidad & Tobago

Conservator of Forests
Forestry Division
Private Bag 30, Long Circular Road
PORT—OF—SPAIN
Trinidad and Tobago

Head
(Wildlife Section) Forestry Division
Ministry of the Environment and National Service
PO Box 30 Tel: (809) 622 3217/622 4521
ST JAMES
Trinidad

Tunisia

Service des parcs nationaux
Ministère de l'agriculture
Direction générale des forêts
30 rue Alain Savary Tel: (216) 1 282681
TUNIS Tlx: 13378
Tunisia

Turkey

Tarim Orman ve Köyisleri Bakanligi
Orman Genel Müdürlügü
(Director General of Forests)
Milli Parklar Dairesi
(National Parks Department)
11 No. lu Bina
Gazi
ANKARA — Tlx: 42108 OGMD TR
Turkey

Turks & Caicos

Director
Department of National Heritage & Parks
Ministry of Natural Resources
GRAND TURK — Tel: (1) 809 946 2855
Turks and Caicos Islands — Tlx: 8227 CHIEFMIN TQ
West Indies — FAX: (1) 809 946 2448

Tuvalu

Secretary
Ministry of Commerce & Natural Resources
VAIAKU — FAX: (688) 826
Funafuti Island
Tuvalu

Uganda

Chief Game Warden
Game Department
P O Box 4
ENTEBBE — Tel: (256) 42 20073
Uganda

Director
Uganda National Parks
Plot 107
6th Street
Industrial Area
P O Box 3530
KAMPALA
Uganda

Chief Environment Officer
Department of Environment Protection
Ministry of Energy, Minerals & Environment Protection
P O Box 9629 Tel: (256) 42 257976
KAMPALA FAX: (256) 42 244801
Uganda

Ukraine

Environment Protection Ministry
5 Chreschatyk
252 001 KIEV Tel: (7) 44 226 2430
Ukraine FAX: (7) 44 229 8383

United Arab Emirates

Ministry of Agriculture &
Fisheries
P O Box 213 Tel: (971) 2 662781
ABU DHABI
United Arab Emirates

Ministry of Agriculture and Fisheries
P O Box 1509
DUBAI
United Arab Emirates

United Kingdom

Director
Countryside Commission
John Dower House
Crescent Place
CHELTENHAM
Gloucestershire GL50 3RA — Tel: 0242 521381
United Kingdom — FAX: (44) 242 584270

Director
Countryside Commission for Scotland
Battleby
Redgorton
PERTH PH1 3EW — Tel: (44) 738 27921
Scotland, United Kingdom — FAX: (44) 738 30583

English Nature
Northminster House — Tel: (44) 733 40345
PETERBOROUGH PE1 1UA — Tlx: 2130132
United Kingdom — FAX: (44) 733 898290

Joint Nature Conservation Committee
Monkston House
PETERBOROUGH PE1 1JY — Tel: (44) 733 62626
United Kingdom — FAX: (44) 733 555948

Nature Conservancy Council for Scotland
12 Hope Terrace
EDINBURGH EH9 2AS — Tel: (44) 31 447 4784
Scotland — FAX: (44) 31 447 0055
United Kingdom

Countryside Council for Wales
Plas Penrhos
Ffordd Penrhos
BANGOR — Tel: (44) 248 370444
Gwynedd LL57 2LQ — FAX: (44) 248 355782
Wales, United Kingdom

Department of the Environment
(Northern Ireland)
Calvert House
23 Castle Place
BELFAST BT1 1FY — Tel: (44) 232 63210
Northern Ireland, United Kingdom

Forestry Commission
231 Corstorphine Road
EDINBURGH EH12 7AT — Tel: (44) 31 344 0303
Scotland

Department of Local Government & the Environment
Central Government Offices
DOUGLAS
Isle of Man
United Kingdom

(Antarctic – Tristan da Cunha & Gough Island)
Secretary
Foreign & Commonwealth Office
West Indian and Atlantic Department
LONDON SW1 2AH — Tel: (44) 71 233 7810
United Kingdom

(Antarctic – Falkland Islands, South Georgia & South Sandwich Islands)
The Chief Executive
The Secretary
STANLEY
Falkland Islands

United States of America

National Park Service
Department of the Interior — Tel: (1) 202 343 7063
P O Box 37127 — Tlx: 1561375
WASHINGTON DC 20013—7127 — FAX: (1) 202 343 7059
USA

Fish & Wildlife Service
Department of the Interior
WASHINGTON DC 20240 — Tel: (1) 202 343 5634
USA

Forest Service
Department of Agriculture
P O Box 2417 — Tel: (1) 202 447 3760
WASHINGTON DC 20013
USA

Bureau of Land Management
Department of the Interior
18th and C Streets NW — Tel: (1) 202 343 9435
WASHINGTON DC 20240
USA

Director
Division of State Parks
Department of Conservation & Natural Resources
64 N Union Street
MONTGOMERY Tel: (1) 205 242 3486
Alabama 36130
USA

Director
Divison of Parks & Outdoor Recreation
Department of Natural Resources
400 Willoughby
JUNEAU Tel: (1) 907 465 2400
Alaska 99801
USA

Director
State Parks Board
Natural Resources Division
Land Department
1616 W Adams St
PHOENIX Tel: (1) 602 542 4621
Arizona 85007
USA

Director
Parks Division
Department of Parks & Tourism
One Capital Mall
Capitol
LITTLE ROCK Tel: (1) 501 682 2345
Arkansas 72201, USA

Director
Department of Parks & Recreation
The Environmental Affairs Agency
1416 Ninth Street
P O Box 942896
SACRAMENTO Tel: (1) 916 445 2358
California 94296—0001
USA

Director
Division of Parks & Outdoor Recreation
Department of Natural Resources
1313 Sherman, Rm 718
DENVER Tel: (1) 303 866 3437
Colorado 80203
USA

Chief
Bureau of Parks & Forests
Department of Environmental Protection
State Office Building
165 Capitol Ave
HARTFORD Tel: (1) 203 566 2287
Connecticut 06106
USA

Director
Division of Parks & Recreation
Department of Natural Resources
& Environmental Control
89 Kings Highway
P O Box 1401
DOVER Tel: (1) 302 739 4401
Delaware 19903
USA

Director
Division of Recreation & Parks
Department of Natural Resources
Marjory Stoneman Douglas Bldg.
TALLAHASSEE — Tel: (1) 904 488 6131
Florida 32399
USA

Director
Parks & Recreation/Historic Sites Division
Department of Natural Resources
Floyd Towers East
205 Butler St
ATLANTA — Tel: (1) 404 656 3530
Georgia 30334
USA

Department of Parks and Recreation
490 Chalan Palasyu
AGANA HEIGHTS
Guam
USA 96910

Director
Division of State Parks & Recreation
Department of Land & Natural Resources
Box 621
HONOLULU — Tel: (1) 808 548 7455
Hawaii 96809
USA

National Park Service

300 Ala Moana Blvd
HONOLULU — Tel: (1) 808 541 2693
HI 96850 — FAX: (1) 808 541 3696
USA

Chairman
Idaho State Parks & Recreation
Statehouse Mail
BOISE — Tel: (1) 208 334 2154
Idaho 83720
USA

Chief
Nature Preserves System
Department of Conservation
Lincoln Tower Plaza
524 S Second St
SPRINGFIELD — Tel: (1) 217 782 6302
Illinois 62701—1787
USA

Head
Division of State Parks
Department of Natural Resources
608 State Office Bldg
INDIANAPOLIS — Tel: (1) 317 232 4124
Indiana 46204
USA

Administrator
Parks, Recreation & Preserves Division
Department of Natural Resources
E Ninth & Grand Ave
Wallace Bldg
DES MOINES — Tel: (1) 515 281 5145
Iowa 50319—0034
USA

Secretary
Kansas Department of Wildlife & Parks
900 Jackson St, # 502
TOPEKA — Tel: (1) 913 296 2281
Kansas 66612—1220
USA

Commissioner of Parks
Department of Parks
Capital Plaza Bldg, 10th floor
FRANKFORT — Tel: (1) 502 564 2172
Kentucky 40601
USA

Director
Office of State Parks
Department of Culture, Recreation & Tourism
P O Box 44426
BATON ROUGE — Tel: (1) 504 342 8111
Louisiana 70804
USA

Director
Bureau of Parks & Recreation
Department of Conservation
State House Station #22
AUGUSTA — Tel: (1) 207 289 3821
Maine 04333
USA

Director
State Parks & Forests
Forest, Park & Wildlife Services
Department of Natural Resources
Tawes State Office Bldg
ANNAPOLIS — Tel: (1) 301 974 3776
Maryland 21401
USA

Director
Division of Forests & Parks
Department of Environmental Management
Executive Office of Environmental Affairs
Leverett Saltonstall Bldg
100 Cambridge St
BOSTON — Tel: (1) 617 727 3180
Massachusetts 02202
USA

Chief
Parks Division
Department of Natural Resources
Box 30028
LANSING — Tel: (1) 517 373 1270
Michigan 48909
USA

Director
Division of Parks & Recreation
500 Lafayette Rd
ST PAUL — Tel: (1) 612 296 2270
Minnesota 55155—4001
USA

Executive Director
Division of Parks & Recreation
Department of Wildlife, Fisheries & Parks
P O Box 23093
JACKSON — Tel: (1) 601 961 5300
Mississippi 39225
USA

Director
Division of Parks, Recreation & Historic Preservation
P O Box 176
JEFFERSON CITY — Tel: (1) 314 751 2479
Missouri 65102
USA

Administrator
Parks
Department of Fish, Wildlife & Parks
1420 East Sixth
HELENA — Tel: (1) 406 444 3750
Montana 59620
USA

Chief
State Parks
Game & Parks Commission
2200 N 33rd St
P O Box 30370
LINCOLN — Tel: (1) 402 471 5511
Nebraska 68503
USA

Administrator
Division of State Parks
Department of Conservation & Natural Resources
350 Capitol Hill Ave
P O Box 11100
RENO Tel: (1) 702 687 438
Nevada
USA

Director
Division of Parks
Department of Resources & Economic Development
P O Box 856
1105 Loudon Road
CONCORD Tel: (1) 603 271 3254
New Hampshire 03301
USA

Director
Division of Parks & Forestry
Department of Environmental Protection
CN 404
TRENTON Tel: (1) 609 292 2733
New Jersey
USA

Director
Parks & Recreation Division
Energy, Minerals & Natural Resources Department
Villagra Building
P O Box 1147
SANTA FE Tel: (1) 505 827 5906
New Mexico 87504—1147
USA

Commissioner
State Office of Parks, Recreation &
Historic Preservation
Empire State Plaza
ALBANY Tel: (1) 518 474 0456
New York 12238
USA

Director
State Parks & Recreation
Department of Environmental Health &
Natural Resources
P O Box 27687
RALEIGH Tel: (1) 919 733 4181
North Carolina 27603
USA

Executive Director
Parks & Recreation Department
1424 W Century Ave, # 202
BISMARCK Tel: (1) 701 224 4887
North Dakota 58501
USA

Chief
Division of Parks & Recreation
Department of Natural Resources
Fountain Square
COLUMBUS Tel: (1) 614 265 6610
Ohio 43224

Director
Division of State Parks
Tourism & Recreation Department
500 Will Rogers Memorial Bldg
OKLAHOMA CITY Tel: (1) 405 521 3411
Oklahoma 73105
USA

Chief
Parks and Recreation Branch
Department of Transportation
324 Capitol Street NE
SALEM Tel: (1) 503 378 8486
Oregon 97310
USA

Director
Bureau of State Parks
Department of Environmental Resources
Fulton Bldg
P O Box 2063
HARRISBURG Tel: (1) 717 787 2703
Pennsylvania 17120
USA

Director
National Park Service
P O Box 712
OLD SAN JUAN
Puerto Rico 00902

Chief
Division of Parks & Recreation
Department of Environmental Management
9 Hayes St
PROVIDENCE — Tel: (1) 401 277 2635
Rhode Island 02908
USA

Executive Director
Department of Parks, Recreation & Tourism
Edgar A Brown Bldg
1205 Pendleton St
COLUMBIA
South Carolina 29201
USA

Department Secretary
Game, Fish & Parks Department
445 East Capitol
PIERRE — Tel: (1) 605 773 3381
South Dakota 57501—3185
USA

Director
Division of Parks & Recreation
Department of Conservation
701 Broadway
Customs House
NASHVILLE — Tel: (1) 615 742 6745
Tennessee 37234—0435
USA

Director
Parks
Parks & Wildlife Department
4200 Smith School Road
AUSTIN
Texas 78744
USA
Tel: (1) 512 389 4866

Director
Division of Parks & Recreation
State Department of Natural Resources
1636 W North Temple
SALT LAKE CITY
Utah 84116—3156
USA
Tel: (1) 801 538 7220

Director of Parks
Department of Forests, Parks & Recreation
Agency of Natural Resources
10 South
WATERBURY
Vermont 05677
USA
Tel: (1) 802 244 8715

Commissioner
Division of State Parks
Department of Conservation & Recreation
203 Governor St, #306
RICHMOND
Virginia 23219
USA
Tel: (1) 804 786 2132

Deputy Commissioner
Parks & Recreation Department
Conservation & Cultural Affairs
P O Box 4339
ST THOMAS
US Virgin Islands

Director
State Parks & Recreation Commission
7150 Cleanwater Ln
OLYMPIA — Tel: (1) 206 753 5757
Washington 98504—5711
USA

Director
Division of Parks & Recreation
Department of Natural Resources
1900 Kanawha Blvd East
CHARLESTON — Tel: (1) 304 348 2754
West Virginia 25305
USA

Director
Bureau of Parks & Recreation
Department of Natural Resources
Box 7921
MADISON — Tel: (1) 608 266 2185
Wisconsin 53707
USA

Chief
Wyoming State Parks & Historic Sites
2301 Central Ave
Barrett Bldg
CHEYENNE — Tel: (1) 307 777 6025
Wyoming 82002
USA

Uruguay

Director General
Direccion General de Recursos Naturales Renovables
Ministerio de Ganaderia, Agricultura y Pesca
Cerrito 322, 2° piso
MONTEVIDEO
Uruguay
Tel: (598) 2 959878
FAX: (598) 2 956456

Russian Federation

Ministry of the Ecology & Natural Resources
Nezhdanova ul. 11
103 009 MOSCOW
Russian Federation
Tel: (7) 95 229 5759/281 9004
Tlx: 411692 borei su
FAX: (7) 95 229 6407/230 2792

Vanuatu

Environmental Unit
Ministry of Lands, Geology, Minerals and Rural Water Supply
PMB 007
PORT VILA
Vanuatu
Tel: (678) 3105
Tlx: 1040 VANGOV UH
FAX: (678) 23142

Environmental Unit
Department of Physical Planning & Environment
Ministry of Home Affairs
PMB 036
PORT VILA
Vanuatu
Tel: (678) 22252 x 38
Tlx: 1040 vangov uh
FAX: (678) 23142

Venezuela

Director General Sectoral
MARNR/ODEPRIC/DCAI
Apartado 6623
CARACAS 1010—A
Venezuela
Tel: (58) 2 408 1107
FAX: (58) 2 483 2445

Director
Instituto de Parques (INPARQUES)
Apartado 76471
CARACAS 1070—A, Venezuela
Tel: (58) 2 239 0901
Tlx: 24362 inap vc
FAX: (58) 02 285 3070/239 2698

Viet Nam

Protected Areas Division
123 Lo Duc
HANOI
Socialist Republic of Viet Nam

Western Samoa

National Parks and Reserves Officer
Department of Agriculture, Forests & Fisheries
PO Box L1874
APIA
Western Samoa
Tel: (685) 224812
Tlx: MALO 221 SX
FAX: (685) 21504

Yemen

General Directorate of Forest & Range
Ministry of Agriculture and Fisheries
Al Ziran St
SANA'A
Yemen Arab Republic

Ministry of Culture & Tourism
Aden
Yemen People's Democratic Republic

Department of Forestry
Agricultural Resources Centre
Ministry of Agriculture and Agrarian Reform
PO Box 1161
Aden
Yemen People's Democratic Republic

National Environment Council
Aden
Yemen People's Democratic Republic

Yugoslavia

Zavod SR Slovenije za varstvo naravne in kulturne dediscine
Plecnikov trg. 2, p.p. 176
61001 LJUBLJANA Tel: (38) 61 213012/213083
Slovenia FAX: (38) 61 213120
Yugoslavia

Republicki zavod za zastitu prirode
Ilica 44/11
41000 ZAGREB Tel: (38) 41 432022/23
Croatia
Yugoslavia

Republicki zavod za zastitu prirode
Trg. Nikole Kovacevica 7
P O Box 2
81000 TITOGRAD Tel: (38) 81 22992
Montenegro, Yugoslavia

Zavod za zastitu spomenika kulkure, prirodnih
znamenitosti i rijetkosti
Obala 27, jula 11a, p.p. 650
71001 SARAJEVO Tel: (38) 71 611565
Bosnia and Herzegovina 643555
Yugoslavia

Republicki zavod za zastitu na prirodna
i prirodni retkosti
Evlia Celebija b.b.
91000 SKOPJE Tel: (38) 91 251133
Macedonia
Yugoslavia

Repulicki zavod za zastitu prirode
Treci Bulevar 106, p.f. 51
11070 BEOGRAD — Tel: (38) 11 142 281/165
Serbia
Yugoslavia

Pokrajinski zavod za zastitu prirode
Petrovaradinska tvrdjava
21000 NOVI SAD/Petrovasadin — Tel: (38) 21 432200
Vojvodina
Serbia
Yugoslavia

Pokrajinski zavod zastitu prirode
Lenjinova 13
38000 PRISTINA — Tel: (38) 38 27027
Kosovo
Serbia
Yugoslavia

Zaire

Président délégué général
Institut zaïrois pour la conservation
de la nature (IZCN)
B P 868 — Tel: (243) 12 32668/33250
KINSHASA I — Tlx: 21112
Zaïre — FAX: (243) 12 27547

Directeur général des forêts
Departement de l'environnement, de la conservation
de la nature et du tourisme
B P 73 — Tel: (243) 12 30250
KINSHASA I
Zaïre

Zambia

Director
National Parks and Wildlife Service
Private Bag 1
CHILANGA
Zambia

Tel: (260) 278524
FAX: (260) 278113

Zimbabwe

Director
Department of National Parks
and Wild Life Management
P O Box 8365
Causeway
HARARE
Zimbabwe

Tel: (263) 4 707624/792783
FAX: (263) 4 724914

Bibliography

The following contains both references from the text and additional references relevant to protected landscapes.

Aitchison, J.W. (1987) *The National and Regional Parks of France.* University College of Wales, Aberystwyth.

Barber, Sir D. (1988) Foreword, *Protected Landscapes: Summary Proceedings of an International Symposium, Lake District, United Kingdom* (ed. J. Foster), IUCN, Gland, Switzerland.

Batisse, M. (1986) Developing and focussing the biosphere reserve concept. *Nature and Resources,* 22 July–September.

Council of Europe Committee of Ministers (1973) Resolution (73)30 on the European Terminology for Protected Areas. Division of Environment and Natural Resources.

Countryside Commission (1987a) *Landscape Assessment: A Countryside Commission Approach.* Countryside Commission, Cheltenham, UK.

Countryside Commission (1987b) *The Lake District Declaration.* Countryside Commission, Cheltenham, UK.

Countryside Commission (1988) *National Parks in England and Wales* (brochure). Countryside Commission, Cheltenham, UK.

Dasmann, R.F. (1973) *Classification and Use of Protected Natural and Cultural Areas.* IUCN Occasional Paper No. 4, IUCN, Morges, Switzerland.

Department of Conservation (1988) *Protected Areas Legislation Review: Issues for public comment.* Wellington, New Zealand.

Department of Conservation (1990) *New Zealand Coastal Policy Statement.* Wellington, New Zealand.

Diamont, J., Eugster, G. and Duerksen, C.J. (1984) *A Citizen's Guide to River Conservation.* The Conservation Foundation, Washington, DC.

Dower, J. (1945) *National Parks in England and Wales.* Ministry of Town and Country Planning, UK.

Dubos, R. (1980) *The Wooing of Earth.* Charles Scribner's Sons, New York.

Dumeige, B., Plummer, B. and Beresford, M. (1988) *Harnessing Resources in the Normandie Maine Regional Park.* Countryside Commission. Cheltenham, UK.

Eidsvik, H.K. (1987) Closing address to International Symposium. In: *Protected Landscapes: Summary Proceedings of an International Symposium, Lake District, United Kingdom,* (ed. J. Foster), IUCN, Gland, Switzerland.

English Tourist Board and Countryside Commission (1990) *Principles for Tourism in National Parks*. London and Cheltenham.
European Parliament Committee on Regional Policy and Regional Planning (1988) *The Creation of Parks, the Protection of Land and the Development of Farm Holidays*, (Draft report).
Forestry Council (1980) *Policy Guidelines for Private Indigenous Forests*. Wellington, New Zealand.
Forman, R.T.T. (ed.) (1979) *Pine Barrens: Ecosystem and Landscape*. Academic Press, New York.
Forman, R.T.T. and Godron, M. (1986) *Landscape Ecology*. John Wiley and Sons, New York.
Forster, M.J. (1987) *Protected Landscapes – The Legislative Background*. IUCN Environmental Law Centre, Bonn, Germany.
Foster, J. (ed.) (1988) *Protected Landscapes: Summary Proceedings of an International Symposium, Lake District, United Kingdom*. IUCN, Gland, Switzerland.
Harrison, J. and Karpowicz, Z. (1987) *Protected Landscapes – Facts and Figures*. Protected Areas Data Unit, IUCN Conservation Monitoring Centre, Cambridge, UK.
Holdgate, M.W. (1989) National Parks in a Changing World. Address to 40th Anniversary celebration of UK national parks. IUCN, Gland, Switzerland.
International Federation of Landscape Architects information leaflet. (1973) Amsterdam.
IUCN (1978) *Categories, Objectives and Criteria for Protected Areas*. IUCN, Morges, Switzerland.
IUCN (1980) *World Conservation Strategy: Living Resource Conservation for Sustainable Development*. IUCN–UNEP–WWF, Gland, Switzerland.
IUCN (1985) *1985 United Nations List of National Parks and Protected Areas*. IUCN, Gland, Switzerland.
IUCN (1990a) *1990 United Nations List of National Parks and Protected Areas*. IUCN, Gland, Switzerland.
IUCN (1990b) *A Conservation and Development Strategy for Ngorongoro*: Final conclusions and recommendations of a Workshop held at Dar-es-Salaam, December 1989. IUCN Regional Office Eastern Africa, Nairobi, Kenya.
IUCN (1991) *Caring for the Earth: A Strategy for Sustainable Living*. IUCN–UNEP–WWF, Gland, Switzerland.

IUCN Conservation. Monitoring Centre (1987) *Protected Landscapes: Experience Around the World.* IUCN, Gland, Switzerland and Cambridge, UK.

Kayera, J.A. (1988) *Balancing Conservation and Human Needs in Tanzania: The Case of Ngorongoro* (unpublished).

Kelleher, G. and Kenchington, R.A. (1982) Australia's Great Barrier Reef Marine Park: Making Development Compatible with Conservation. *Ambio* 11 (5).

Kelleher, G. (1987) *Marine Protected Areas. Proceedings of First Annual Conference,* Environmental Institute of Australia, Sydney.

Kelleher, G., Childs, R. and Quilty, P. (1989) *Great Barrier Reef Area: A Protected Seascape.* Great Barrier Reef Marine Park Authority. Canberra, ACT, Australia.

Lucas, D. (1987) *Contemporary Rural Planning Theory.* Geraldine, New Zealand.

McAfee, P. (1988) *Environment and Local Authority Planning: A Survey of Environmental Factors in Selected District Schemes.* Works Consultancy Services, Christchurch, New Zealand.

MacEwen, A. and MacEwen, M. (1987) *Greenprints for the Countryside?* Allen and Unwin, London.

MacKinnon, J.R., MacKinnon, K., Child, G. and Thorsell, J. (1986) *Managing Protected Areas in the Tropics.* IUCN, Gland, Switzerland.

McNeely, J.A. (1988) *Economics and Biological Diversity: Developing and Using Economic Incentives to Conserve Biological Diversity.* IUCN, Gland, Switzerland.

McNeely, J.A. and Miller, K.R. (1983) IUCN, National Parks, and Protected Areas: Priorities for Action. *Environmental Conservation,* 10 (1).

McNeely, J.A. and Miller, K.R. (eds) (1984) *National Parks, Conservation, and Development: The Role of Protected Areas in Sustaining Society.* Smithsonian Institution Press, Washington, DC.

McNeely, J.A., Miller, K.R., Reid, W.V., Mittermeier, R.A. and Werner, T.B. (1990) *Conserving the World's Biological Diversity.* IUCN, WRI, GI, WWF–US, the World Bank, Gland, Switzerland, and Washington, DC.

McPhee, J. (1967) *The Pine Barrens.* Farrar, Straus and Giroux, New York, USA.

Mercer, I. (1984) *Dartmoor, in Britain's National Parks* (ed. William Lacey), Windward, England.

Nature Conservation Bureau (1985) *Nature Conservation Administration in Japan*. Nature Conservation Bureau, Environment Agency, Government of Japan, Tokyo, Japan.

Ministry of Urban and Rural Reconstruction and Environment Protection (1987) *Mount Taishan: Nomination for World Heritage Listing*. Beijing, People's Republic of China.

Peak Park District Planning Board (1988) *Peak National Park Plan*, Draft First Review. Bakewell, Derbyshire, UK.

Phillips, A. (1985) *Chairman's and Director's Visit to Vercors Regional Nature Park, France*. Countryside Commission Information Bulletin, Cheltenham, UK.

Phillips, A. (1988) Landscape conservation: British experience and world conservation needs: Paper for the British Association for the Advancement of Science.

Plummer, B. (1983) *French Regional Nature Parks: An Example of Changing Perspectives in Countryside Management*. University of Bristol, UK.

Poore, D. and Poore, J. (1987) *Protected Landscapes: The United Kingdom Experience*. IUCN, Gland, Switzerland.

Poore, D. and Sayer, J. (1987) *The Management of Tropical Moist Forest Lands: Ecological Guidelines*. IUCN, Gland, Switzerland.

Queen Elizabeth II National Trust (1983) *The Waipa County Landscape* No. 1: *An Introduction*. Wellington, New Zealand.

Robertson Vernhes, J.R. (1987) The Work of UNESCO Related to Protected Landscapes, *Protected Landscapes: Summary Proceedings of an International Symposium, Lake District, United Kingdom* (ed. J. Foster), IUCN, Gland, Switzerland.

Sherpa, M.N. (1987) *People, Park Problems and Challenges in the Annapurna Conservation Area in Nepal*. The King Mahendra Trust for Nature Conservation, Kathmandu, Nepal.

Sherpa, M.N., Coburn, B. and Gurung, C.P. (1986) *Annapurna Conservation Area, Nepal: Operational Plan*. The King Mahendra Trust for Nature Conservation, Kathmandu, Nepal.

Simpson, P., Evans, B. and Geden, B. (1987) *Ecological Districts as a Framework for Landscape Interpretation for Tourism in New Zealand*. The Landscape, New Zealand, Summer/Autumn.

Thom, D.A. (1987) *Heritage: The Parks of the People.* Lansdowne Press, Auckland, New Zealand.

Udvardy, M. (1975) *A Classification of the Biogeographic Provinces of the World.* IUCN Occasional Paper No. 18, IUCN, Switzerland.

UNESCO (1972) Convention Concerning the Protection of the World Cultural and Natural Heritage. Paris.

UNESCO (1985) Action plan for biosphere reserves. *Environmental Conservation,* 12, 17–27.

US Department of the Interior, National Park Service (1983) *Greenway Planning: A Conservation Strategy for Significant Landscapes.* Philadelphia, PA.

World Bank (1988) *Wildlands: Their Protection and Management in Economic Development.* World Bank. Washington, DC, 278 pp.

World Commission on Environment and Development (WCED) (1987) *Our Common Future.* Oxford University Press, Oxford, UK.

World Wide Fund For Nature (WWF) and Overseas Development Natural Resources Institute (ODNRI) (1989) *Cross River National Park (Oban Division): Plan for Developing the Park and its Support Zone.* Godalming, Surrey and Chatham, Kent, UK.

Index

Plate page numbers are prefaced by 'pl.'